ESSENTIALS OF STATISTICS

Edited by
Maria Catherine C. Borres

ArclerPress

Essentials of Statistics

Edited by Maria Catherine C. Borres

Edition 2018

ArclerPress

Published by
Arcler Press LLC
708 3rd Avenue, 6th Floor
New York
NY 10017
United States of America
www.arclerpress.com

ISBN: 978-1-68094-585-0 (Hardback)

Printed in United States of America

Library of Congress Control Number: 2017933174

About the Editor

Maria Catherine C. Borres

Catherine is currently taking up Master of Arts in Education Major in Mathematics in Philippine Normal University – Manila. She is currently working as a Content Developer for Mathematics at the Affordable Private Education Center (APEC Schools).

Table of Contents

List of Abbreviations xi
Preface xiii

1 INTRODUCTION TO STATISTICS

Introduction 1
Scope 4
Mathematical Statistics 4
Overview 8
Data Collection 9
Sampling 9
Experimental and Observational Studies 9
Experiments 10
Observational Study 10
Types of Data 11
Statistical Data Type 11
Terminology And Theory Of Inferential Statistics 16
Statistics, Estimators and Pivotal Quantities 16
Null Hypothesis and Alternative Hypothesis 16
Error 17
Interval Estimation 18
Significance 19
Misinterpretation: Correlation 22
History of Statistical Science 22
Applications 24
Applied Statistics, Theoretical Statistics and Mathematical Statistics 24
Machine Learning and Data Mining 24
Statistics in Society 25
Statistical Computing 25
Statistics Applied to Mathematics or the Arts 26

References....26

2 DESCRIPTIVE STATISTICS

Introduction....31
Use In Statistical Analysis....32
Univariate Analysis....32
Bivariate and Multivariate Analysis....32
Continuous Data....33
Central Tendency....33
Medians for Samples....47
Estimation of Variance from Sample Data....50
Statistical Dispersion....60
Shape of a Probability Distribution....62
Descriptions of Shape....63
References....64

3 DATA COLLECTION

Introduction....69
Importance....70
Types....70
Impact of Faulty Data....71
Study Design....71
Statistical Population....71
Effect Size....73
Odds Ratio....82
Risk Difference....83
Rank-Biserial Correlation....84
Effect Size for Ordinal Data....85
t-test for Mean Difference of Single Group or Two Related Groups....85
Statistical Power....88
Sample Size Determination....92
Tables....94
Mead's Resource Equation....95
Missing Data....98
Missing at Random....99
Survey Methodology....102
Sampling....102
Simple Random Sampling....105
Stratified Sampling....108

Minimax Sampling 112
Accidental Sampling 112
Line-Intercept Sampling 113
Panel Sampling 113
Snowball Sampling 113
Theoretical Sampling 114
Standard Error 116
Larger Sample Sizes Give Smaller Standard Errors 118
References 124

4 STATISTICAL INFERENCE

Introduction 129
Models and Assumptions 130
Paradigms for Inference 133
Statistical Assumption 135
Classes of Assumptions 136
Checking Assumptions 136
Estimation Theory 137
Basics 137
Statistical Hypothesis Testing 138
Variations and Sub-Classes 138
The Testing Process 139
Radioactive Suitcase 145
Size 147
Common Test Statistics 149
Origins and Early Controversy 151
Versus Null Hypothesis Statistical Significance Testing 154
References 158

5 CORRELATION AND DEPENDENCE

Introduction 167
Pearson's Product-Moment Coefficient 168
Rank Correlation Coefficients 170
Other Measures of Dependence among Random Variables 171
Sensitivity to the Data Distribution 171
Correlation Matrices 173
Common Misconceptions 173
Bivariate Normal Distribution 175
Partial Correlation 175

Pearson Product-Moment Correlation Coefficient 175
Definition 176
Mathematical Properties 179
Interpretation 179
Inference 182
Pearson's Correlation and Least Squares Regression Analysis 185
Sensitivity to the Data Distribution 186
Variants 187
Heavy Noise Conditions 190
Missing Data 190
Removing Correlation 191
Confounding 191
Causal Definition 192
Control of Confounding 192
History 194
Types of Confounding 194
Decreasing the Potential for Confounding 196
Coefficient of Determination 198
Definitions 199
Interpretation 202
Adjusted R^2 204
Coefficient of Partial Determination 205
Generalized R^2 206
Comparison with Norm of Residuals 206
References 207

6 REGRESSION ANALYSIS

Introduction 211
History 213
Necessary Number of Independent Measurements 214
Statistical Assumptions 215
Underlying Assumptions 215
Linear Regression 216
General Linear Model 218
Diagnostics 218
Interpolation and Extrapolation 219
Nonlinear Regression 219
Power and Sample Size Calculations 220
Other Methods 220
Software 220
Local Regression 221

Definition of a LOESS Model 221
Localized Subsets of Data 222
Degree of Local Polynomials 222
Weight Function 223
Advantages of LOESS 223
Disadvantages of LOESS 223
Robust Regression 224
Applications 224
History and Unpopularity of Robust Regression 225
Methods for Robust Regression 226
Example: BUPA Liver Data 227
Segmented Regression 229
Segmented Linear Regression, Two Segments 229
Example 231
Test Procedures 232
No-effect range 234
Stepwise Regression 235
Main Approaches 235
Selection Criterion 236
Model Accuracy 236
Criticism 237
References 237

7 CATEGORICAL VARIABLE

Introduction 241
Examples of Categorical Variables 242
Notation 242
Number of Possible Values 243
Categorical Variables And Regression 244
Dummy Coding 244
Effects Coding 245
Contrast Coding 246
Nonsense Coding 247
Interactions 247
Categorical by Categorical Variable Interactions 247
Categorical by Continuous Variable Interactions 247
References 248

8 SURVIVAL ANALYSIS

Introduction 249

Introduction to Survival Analysis 250
Definitions of Common Terms in Survival Analysis 251
Kaplan-Meier Plot for the AML Data 251
Log-rank test: Testing for Differences in Survival in the aml data 254
Cox Proportional Hazards (Ph) Regression Analysis 255
Example: Cox Proportional Hazards Regression Analysis for Melanoma 255
Tree-Structured Survival Models 257
Example Survival Tree Analysis 257
Survival Random Forests 258
General Formulation 258
Lifetime Distribution Function and Event Density 259
Hazard Function and Cumulative Hazard Function 259
Quantities Derived From The Survival Distribution 261
Censoring 261
Fitting Parameters to Data 262
Non-Parametric Estimation 263
Computer Software for Survival Analysis 263
Survival Analysis in R 263
Analyses using the R Package "Survival" 263
Survival Tree Analysis using the rpart Package 265
Survival Random Forest Models Using the RandomForestSRC package 265
References 266

Index 267

List of Abbreviations

ANOVA	Analysis of Variance
CD	Coefficient of Dispersion
FDR	False Discovery Rate
FUV	Fraction of Unexplained Variance
FWER	Family Wise Error Rate
IID	Independent Identically Distributed
IQR	Interquartile Range
MAD	Median Absolute Deviation
MAR	Missing at Random
MCAR	Missing Completely at Random
MDL	Minimum Description Length
MNAR	Missing Not at Random
MSWD	Mean Square Weighted Deviation
NCA	Necessary Condition Analysis
NHST	Null Hypothesis Significance Testing
OR	Odds Ratio
PCFG	Probabilistic Context-Free Grammars
RD	Risk Difference
RR	Relative Risk
RSS	Residual Sum of Squares
SEM	Standard Error of the Mean
SRS	Simple Random Sample

Preface

Introduction

In early days, it was used only of the collection of the information of the population of the state military. But in the modern time it is used in almost all aspects of human related activities. Statistics is the study of numerical information, which is called data. People use statistics as tools to understand information. Learning to understand statistics helps a person react intelligently to statistical claims. Statistics are used in the fields of business, math, economics, accounting, banking, government, astronomy, and the natural and social sciences. Statistics play an important role in business. A successful businessman must be very quick and accurate in decision making. He knows that what his customers wants, he should therefore, know what to produce and sell and in what quantities. Statistics helps businessman to plan production according to the taste of the costumers, the quality of the products can also be checked more efficiently by using statistical methods. Statistics play an important role in economics. Economics largely depends upon statistics. National income accounts are multipurpose indicators for the economists and administrators. Statistical methods are used for preparation of these accounts. In economics research statistical methods are used for collecting and analysis the data and testing hypothesis. Statistical plays a central role in almost all natural and social sciences. The methods of natural sciences are most reliable but conclusions draw from them are only probable, because they are based on incomplete evidence. Statistical helps in describing these measurements more precisely. Statistics play an important role in banking. The banks make use of statistics for a number of purposes. Accounting is impossible without exactness. But for decision making purpose, so much precision is not essential the decision may be taken on the basis of approximation, know as statistics. The correction of the values of current asserts is made on the basis of the purchasing power of money or the current value of it. Statistics plays a vital role in almost all the natural and social sciences.

Content Coverage

Chapter One presents basic concept of statistics such as data types, history of statistics as well as its applications.

Chapters Two and *Three* are descriptive statistics and data collection, disparately. Descriptive statistics are statistics that quantitatively describe or summarize features of a collection of information whereas data collection helps to collect the main points as gathered information.

Chapter Four focuses on statistical inference, the process of deducing properties of an underlying distribution by analysis of data.

Chapters Five and *Six* present about Correlation and Regression analysis, separately. Correlation is any of a broad class of statistical relationships involving dependence, whereas regression analysis is a statistical process for estimating the relationships among variables.

Chapter Seven covers the categorical variable that can take on one of a limited, and usually fixed, number of possible values, assigning each individual or other unit of observation to a particular group or nominal category on the basis of some qualitative property..

Chapter Eight focuses on survival analysis for analyzing the expected duration of time until one or more events happen, such as death in biological organisms and failure in mechanical.

CHAPTER 1

Introduction to Statistics

Chapter Contents

- **Introduction**
- **Scope**
- **Overview**
- **Data Collection**
- **Types Of Data**
- **Terminology and Theory of Inferential Statistics**
- **History of Statistical Science**
- **Applications**

Statistics is the mathematical science involved in the application of quantitative principles to the collection, analysis, and presentation of numerical data. The practice of statistics utilizes data from some population in order to describe it meaningfully, to draw conclusions from it, and make informed decisions. The population may be a community, an organization, a production line, a service counter, or a phenomenon such as the weather. Statisticians determine which quantitative model is correct for a given type of problem and they decide what kinds of data should be collected and examined. Applied statistics concerns the application of the general methodology to particular problems.

INTRODUCTION

Statistics is the study of the collection, analysis, interpretation, presentation, and organization of data. In applying statistics to, e.g., a scientific, industrial, or social problem, it is conventional to begin with a statistical population or a statistical model process to be studied. Populations can be diverse topics such as "all people living in a country" or "every atom composing a crystal". Statistics deals with all aspects of data including the planning of data collection in terms of the design of surveys and experiments.

Some popular definitions are:

- Merriam-Webster dictionary defines statistics as "classified facts representing the conditions of a people in a state – especially the facts that can be stated in numbers or any other tabular or classified arrangement".
- Statistician Sir Arthur Lyon Bowley defines statistics as "Numerical statements of facts in any department of inquiry placed in relation to each other".

When census data cannot be collected, statisticians collect data by developing specific experiment designs and survey samples. Representative sampling assures that inferences and conclusions can safely extend from the sample to the population as a whole. An experimental study involves taking measurements of the system under study, manipulating the system, and then taking additional measurements using the same procedure to determine if the manipulation has modified the values of the measurements. In contrast, an observational study does not involve experimental manipulation.

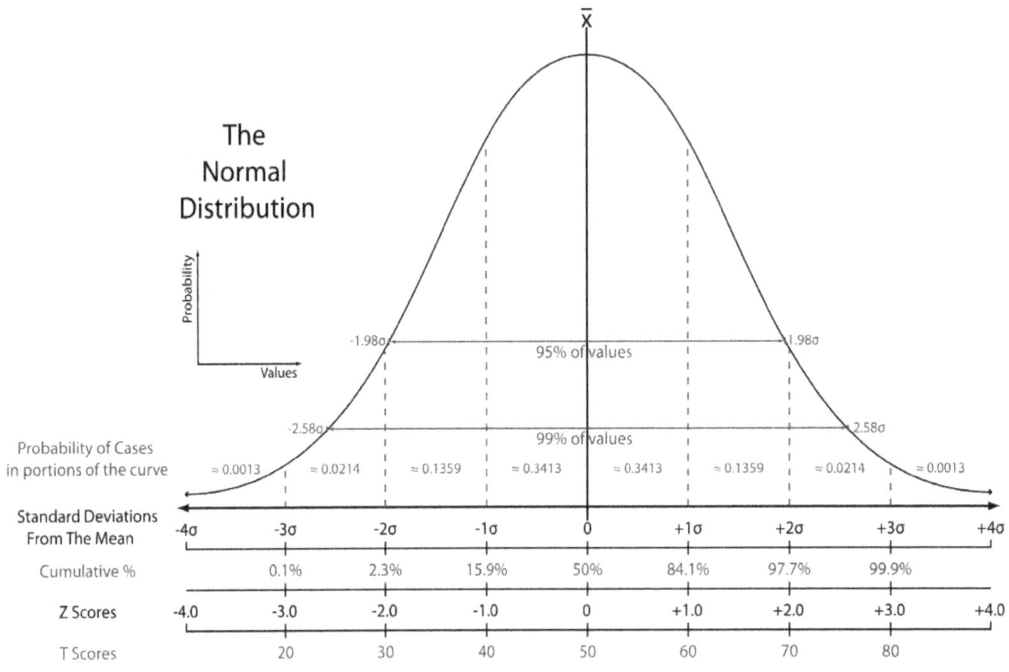

More probability density is found as one gets closer to the expected (mean) value in a normal distribution. Statistics used in standardized testing assessment are shown. The scales include *standard deviations, cumulative percentages, percentile equivalents, Z-scores, T-scores, standard nines,* and *percentages in standard nines.*

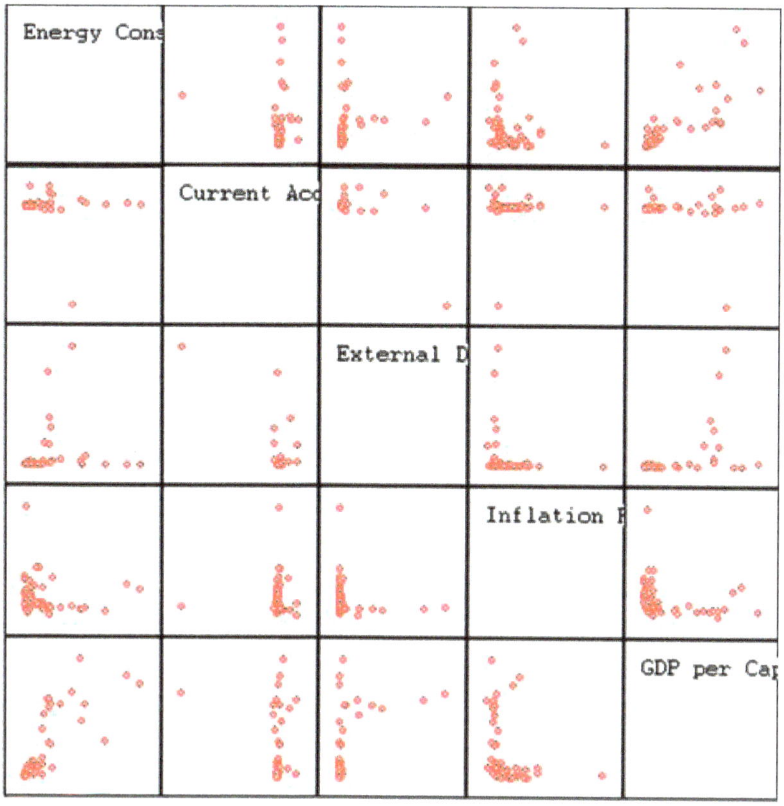

Scatter plots are used in descriptive statistics to show the observed relationships between different variables.

Two main statistical methodologies are used in data analysis: descriptive statistics, which summarizes data from a sample using indexes such as the mean or standard deviation, and inferential statistics, which draws conclusions from data that are subject to random variation (e.g., observational errors, sampling variation). Descriptive statistics are most often concerned with two sets of properties of a *distribution* (sample or population): *central tendency* (or *location*) seeks to characterize the distribution's central or typical value, while *dispersion* (or *variability*) characterizes the extent to which members of the distribution depart from its center and each other. Inferences on mathematical statistics are made under the framework of probability theory, which deals with the analysis of random phenomena.

A standard statistical procedure involves the test of the relationship between two statistical data sets, or a data set and a synthetic data drawn from idealized model. A hypothesis is proposed for the statistical relationship between the two data sets, and this is compared as an alternative to an idealized null hypothesis of no relationship between two data sets. Rejecting or disproving the null hypothesis is done using statistical tests that quantify the sense in which the null can be proven false, given the data that are used in the test. Working from a null hypothesis, two basic forms of error are recognized: Type I errors (null hypothesis is falsely rejected giving a "false positive") and Type II errors (null hypothesis fails to be rejected and an actual difference between populations is missed giving a "false negative"). Multiple problems have come to be associated with this framework: ranging from obtaining a sufficient sample size to specifying an adequate null hypothesis.

Measurement processes that generate statistical data are also subject to error. Many of these errors are classified as random (noise) or systematic (bias), but other types of errors (e.g., blunder, such as when an analyst reports incorrect units) can also be important. The presence of missing data and/or censoring may result in biased estimates and specific techniques have been developed to address these problems.

Statistics can be said to have begun in ancient civilization, going back at least to the 5th century BC, but it was not until the 18th century that it started to draw more heavily from calculus and probability theory. Statistics continues to be an area of active research, for example on the problem of how to analyze Big data.

SCOPE

Statistics is a mathematical body of science that pertains to the collection, analysis, interpretation or explanation, and presentation of data, or as a branch of mathematics. Some consider statistics to be a distinct mathematical science rather than a branch of mathematics. While many scientific investigations make use of data, statistics is concerned with the use of data in the context of uncertainty and decision making in the face of uncertainty.

Mathematical Statistics

Mathematical statistics is the application of mathematics to statistics, which was originally conceived as the science of the state — the collection and analysis of facts about a country: its economy, land, military, population, and so forth. Mathematical techniques used for this include mathematical analysis, linear algebra, stochastic analysis, differential equations, and measure-theoretic probability theory.

Statistical science is concerned with the planning of studies, especially with the design of randomized experiments and with the planning of surveys using random sampling. The initial analysis of the data from properly randomized studies often follows the study protocol.

Of course, the data from a randomized study can be analyzed to consider secondary hypotheses or to suggest new ideas. A secondary analysis of the data from a planned study uses tools from data analysis.

Data analysis is divided into:

- descriptive statistics - the part of statistics that describes data, i.e. summarizes the data and their typical properties.
- inferential statistics - the part of statistics that draws conclusions from data (using some model for the data): For example, inferential statistics involves selecting a model for

the data, checking whether the data fulfill the conditions of a particular model, and with quantifying the involved uncertainty (e.g. using confidence intervals).

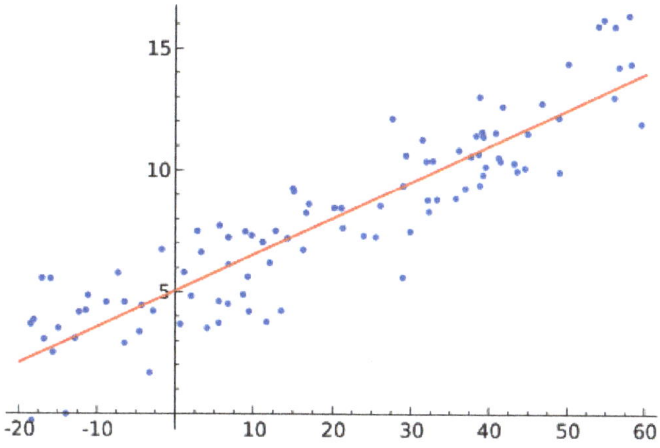

Illustration of linear regression on a data set. Regression analysis is an important part of mathematical statistics.

While the tools of data analysis work best on data from randomized studies, they are also applied to other kinds of data --- for example, from natural experiments and observational studies, in which case the inference is dependent on the model chosen by the statistician, and so subjective.

Mathematical statistics has been inspired by and has extended many options in applied statistics.

Topics

The following are some of the important topics in mathematical statistics:

Probability Distributions

A probability distribution assigns a probability to each measurable subset of the possible outcomes of a random experiment, survey, or procedure of statistical inference. Examples are found in experiments whose sample space is non-numerical, where the distribution would be a categorical distribution; experiments whose sample space is encoded by discrete random variables, where the distribution can be specified by a probability mass function; and experiments with sample spaces encoded by continuous random variables, where the distribution can be specified by a probability density function. More complex experiments, such as those involving stochastic processes defined in continuous time, may demand the use of more general probability measures.

A probability distribution can either be univariate or multivariate. A univariate distribution gives the probabilities of a single random variable taking on various alternative values; a multivariate distribution (a joint probability distribution) gives the probabilities of a random vector—a set of two or more random variables—taking on various combinations of values. Important and commonly encountered univariate probability distributions include the binomial distribution, the hypergeometric distribution, and the normal distribution. The multivariate normal distribution is a commonly encountered multivariate distribution.

Special distributions

- Normal distribution (Gaussian distribution), the most common continuous distribution
- Bernoulli distribution, for the outcome of a single Bernoulli trial (e.g. success/failure, yes/no)
- Binomial distribution, for the number of "positive occurrences" (e.g. successes, yes votes, etc.) given a fixed total number of independent occurrences
- Negative binomial distribution, for binomial-type observations but where the quantity of interest is the number of failures before a given number of successes occurs
- Geometric distribution, for binomial-type observations but where the quantity of interest is the number of failures before the first success; a special c*Discrete uniform distribution, for a finite set of values (e.g. the outcome of a fair die)
- Continuous uniform distribution, for continuously distributed values
- Poisson distribution, for the number of occurrences of a Poisson-type event in a given period of time
- Exponential distribution, for the time before the next Poisson-type event occurs
- Gamma distribution, for the time before the next k Poisson-type events occur
- Chi-squared distribution, the distribution of a sum of squared standard normal variables; useful e.g. for inference regarding the sample variance of normally distributed samples
- Student's t distribution, the distribution of the ratio of a standard normal variable and the square root of a scaled chi squared variable; useful for inference regarding the mean of normally distributed samples with unknown variance
- Beta distribution, for a single probability (real number between 0 and 1); conjugate to the Bernoulli distribution and binomial distribution

Statistical Inferences

Statistical inference is the process of drawing conclusions from data that are subject to random variation, for example, observational errors or sampling variation. Initial requirements of such a system of procedures for inference and induction are that the system should produce reasonable answers when applied to well-defined situations and that it should be general enough to be applied across a range of situations. Inferential statistics are used to test hypotheses and make estimations using sample data. Whereas descriptive statistics describe a sample, inferential statistics infer predictions about a larger population that the sample represents. The outcome of statistical inference may be an answer to the question "what should be done next?", where this might be a decision about making further experiments or surveys, or about drawing a conclusion before implementing some organizational or governmental policy. For the most part, statistical inference makes propositions about populations, using data drawn from the population of interest via some form of random sampling. More generally, data about a random process is obtained from its observed behavior during a finite period of time. Given a parameter or hypothesis about which one wishes to make inference, statistical inference most often uses:

- a statistical model of the random process that is supposed to generate the data, which is known when randomization has been used, and
- a particular realization of the random process; i.e., a set of data.

Regression

In statistics, regression analysis is a statistical process for estimating the relationships among variables. It includes many techniques for modeling and analyzing several variables, when the focus is on the relationship between a dependent variable and one or more independent variables. More specifically, regression analysis helps one understand how the typical value of the dependent variable (or 'criterion variable') changes when any one of the independent variables is varied, while the other independent variables are held fixed. Most commonly, regression analysis estimates the conditional expectation of the dependent variable given the independent variables – that is, the average value of the dependent variable when the independent variables are fixed. Less commonly, the focus is on a quantile, or other location parameter of the conditional distribution of the dependent variable given the independent variables. In all cases, the estimation target is a function of the independent variables called the regression function. In regression analysis, it is also of interest to characterize the variation of the dependent variable around the regression function which can be described by a probability distribution.

Many techniques for carrying out regression analysis have been developed. Familiar methods such as linear regression and ordinary least squares regression are parametric, in that the regression function is defined in terms of a finite number of unknown parameters that are estimated from the data. Nonparametric regression refers to techniques that allow the regression function to lie in a specified set of functions, which may be infinite-dimensional.

Nonparametric Statistics

Nonparametric statistics are statistics not based on parameterized families of probability distributions. They include both descriptive and inferential statistics. The typical parameters are the mean, variance, etc. Unlike parametric statistics, nonparametric statistics make no assumptions about the probability distributions of the variables being assessed.

Non-parametric methods are widely used for studying populations that take on a ranked order (such as movie reviews receiving one to four stars). The use of non-parametric methods may be necessary when data have a ranking but no clear numerical interpretation, such as when assessing preferences. In terms of levels of measurement, non-parametric methods result in "ordinal" data. As non-parametric methods make fewer assumptions, their applicability is much wider than the corresponding parametric methods. In particular, they may be applied in situations where less is known about the application in question. Also, due to the reliance on fewer assumptions, non-parametric methods are more robust.

Another justification for the use of non-parametric methods is simplicity. In certain cases, even when the use of parametric methods is justified, non-parametric methods may be easier to use. Due both to this simplicity and to their greater robustness, non-parametric methods are seen by some statisticians as leaving less room for improper use and misunderstanding.

Statistics, Mathematics, and Mathematical Statistics

Mathematical statistics has substantial overlap with the discipline of statistics. Statistical theorists study and improve statistical procedures with mathematics, and statistical research often raises mathematical questions. Statistical theory relies on probability and decision theory.

Mathematicians and statisticians like Gauss, Laplace, and C. S. Peirce used decision theory with probability distributions and loss functions (or utility functions). The decision-theoretic approach to statistical inference was reinvigorated by Abraham Wald and his successors, and makes extensive use of scientific computing, analysis, and optimization; for the design of experiments, statisticians use algebra and combinatorics.

OVERVIEW

In applying statistics to a problem, it is common practice to start with a population or process to be studied. Populations can be diverse topics such as "all persons living in a country" or "every atom composing a crystal".

Ideally, statisticians compile data about the entire population (an operation called census). This may be organized by governmental statistical institutes. Descriptive statistics can be used to summarize the population data. Numerical descriptors include mean and standard deviation for continuous data types (like income), while frequency and percentage are more useful in terms of describing categorical data (like race).

When a census is not feasible, a chosen subset of the population called a sample is studied. Once a sample that is representative of the population is determined, data is collected for the sample members in an observational or experimental setting. Again, descriptive statistics can be used to summarize the sample data. However, the drawing of the sample has been subject to an element of randomness, hence the established numerical descriptors from the sample are also due to uncertainty. To still draw meaningful conclusions about the entire population, inferential statistics is needed. It uses patterns in the sample data to draw inferences about the population represented, accounting for randomness. These inferences may take the form of: answering yes/no questions about the data (hypothesis testing), estimating numerical characteristics of the data (estimation), describing associations within the data (correlation) and modeling relationships within the data (for example, using regression analysis). Inference can extend to forecasting, prediction and estimation of unobserved values either in or associated with the population being studied; it can include extrapolation and interpolation of time series or spatial data, and can also include data mining.

DATA COLLECTION

Sampling

When full census data cannot be collected, statisticians collect sample data by developing specific experiment designs and survey samples. Statistics itself also provides tools for prediction and forecasting the use of data through statistical models. To use a sample as a guide to an entire population, it is important that it truly represents the overall population. Representative sampling assures that inferences and conclusions can safely extend from the sample to the population as a whole. A major problem lies in determining the extent that the sample chosen is actually representative. Statistics offers methods to estimate and correct for any bias within the sample and data collection procedures. There are also methods of experimental design for experiments that can lessen these issues at the outset of a study, strengthening its capability to discern truths about the population.

Sampling theory is part of the mathematical discipline of probability theory. Probability is used in mathematical statistics to study the sampling distributions of sample statistics and, more generally, the properties of statistical procedures. The use of any statistical method is valid when the system or population under consideration satisfies the assumptions of the method. The difference in point of view between classic probability theory and sampling theory is, roughly, that probability theory starts from the given parameters of a total population to deduce probabilities that pertain to samples. Statistical inference, however, moves in the opposite direction—inductively inferring from samples to the parameters of a larger or total population.

Experimental and Observational Studies

A common goal for a statistical research project is to investigate causality, and in particular to draw a conclusion on the effect of changes in the values of predictors or independent variables on dependent variables. There are two major types of causal statistical studies: experimental studies and observational studies. In both types of studies, the effect of differences of an independent variable (or variables) on the behavior of the dependent variable are observed. The difference between the two types lies in how the study is actually conducted.

Each can be very effective. An experimental study involves taking measurements of the system under study, manipulating the system, and then taking additional measurements using the same procedure to determine if the manipulation has modified the values of the measurements. In contrast, an observational study does not involve experimental manipulation. Instead, data are gathered and correlations between predictors and response are investigated. While the tools of data analysis work best on data from randomized studies, they are also applied to other kinds of data – like natural experiments and observational studies – for which a statistician would use a modified, more structured estimation method (e.g., Difference in differences estimation and instrumental variables, among many others) that produce consistent estimators.

Experiments

The basic steps of a statistical experiment are:

- Planning the research, including finding the number of replicates of the study, using the following information: preliminary estimates regarding the size of treatment effects, alternative hypotheses, and the estimated experimental variability. Consideration of the selection of experimental subjects and the ethics of research is necessary. Statisticians recommend that experiments compare (at least) one new treatment with a standard treatment or control, to allow an unbiased estimate of the difference in treatment effects.
- Design of experiments, using blocking to reduce the influence of confounding variables, and randomized assignment of treatments to subjects to allow unbiased estimates of treatment effects and experimental error. At this stage, the experimenters and statisticians write the *experimental protocol* that will guide the performance of the experiment and which specifies the *primary analysis* of the experimental data.
- Performing the experiment following the experimental protocol and analyzing the data following the experimental protocol.
- Further examining the data set in secondary analyses, to suggest new hypotheses for future study.
- Documenting and presenting the results of the study.

Experiments on human behavior have special concerns. The famous Hawthorne study examined changes to the working environment at the Hawthorne plant of the Western Electric Company. The researchers were interested in determining whether increased illumination would increase the productivity of the assembly line workers. The researchers first measured the productivity in the plant, then modified the illumination in an area of the plant and checked if the changes in illumination affected productivity. It turned out that productivity indeed improved (under the experimental conditions). However, the study is heavily criticized today for errors in experimental procedures, specifically for the lack of a control group and blindness. The Hawthorne effect refers to finding that an outcome (in this case, worker productivity) changed due to observation itself. Those in the Hawthorne study became more productive not because the lighting was changed but because they were being observed.

Observational Study

An example of an observational study is one that explores the association between smoking and lung cancer. This type of study typically uses a survey to collect observations about the area of interest and then performs statistical analysis. In this case, the researchers would collect observations of both smokers and non-smokers, perhaps through a cohort study, and then look for the number of cases of lung cancer in each group. A case-control study is another type of observational study in which people with and without the outcome of interest (e.g. lung cancer) are invited to participate and their exposure histories are collected.

TYPES OF DATA

Various attempts have been made to produce a taxonomy of levels of measurement. The psychophysicist Stanley Smith Stevens defined nominal, ordinal, interval, and ratio scales. Nominal measurements do not have meaningful rank order among values, and permit any one-to-one transformation. Ordinal measurements have imprecise differences between consecutive values, but have a meaningful order to those values, and permit any order-preserving transformation. Interval measurements have meaningful distances between measurements defined, but the zero value is arbitrary (as in the case with longitude and temperature measurements in Celsius or Fahrenheit), and permit any linear transformation. Ratio measurements have both a meaningful zero value and the distances between different measurements defined, and permit any rescaling transformation.

Because variables conforming only to nominal or ordinal measurements cannot be reasonably measured numerically, sometimes they are grouped together as categorical variables, whereas ratio and interval measurements are grouped together as quantitative variables, which can be either discrete or continuous, due to their numerical nature. Such distinctions can often be loosely correlated with data type in computer science, in that dichotomous categorical variables may be represented with the Boolean data type, polytomous categorical variables with arbitrarily assigned integers in the integral data type, and continuous variables with the real data type involving floating point computation. But the mapping of computer science data types to statistical data types depends on which categorization of the latter is being implemented.

Other categorizations have been proposed. For example, Mosteller and Tukey (1977) distinguished grades, ranks, counted fractions, counts, amounts, and balances. Nelder (1990) described continuous counts, continuous ratios, count ratios, and categorical modes of data.

The issue of whether or not it is appropriate to apply different kinds of statistical methods to data obtained from different kinds of measurement procedures is complicated by issues concerning the transformation of variables and the precise interpretation of research questions. "The relationship between the data and what they describe merely reflects the fact that certain kinds of statistical statements may have truth values which are not invariant under some transformations. Whether or not a transformation is sensible to contemplate depends on the question one is trying to answer" (Hand, 2004, p. 82).

Statistical Data Type

In statistics, groups of individual data points may be classified as belonging to any of various statistical data types, e.g. categorical ("red", "blue", "green"), real number (1.68, -5, 1.7e+6), etc. The data type is a fundamental component of the semantic content of the variable, and controls which sorts of probability distributions can logically be used to describe the variable, the permissible operations on the variable, the type of regression analysis used to predict the variable, etc. The concept of data type is similar to the concept of level of measurement, but more specific: For example, count data require a different distribution (e.g. a Poisson distribution or binomial distribution) than non-negative real-valued data require, but both fall under the same level of measurement (a ratio scale).

Various attempts have been made to produce a taxonomy of levels of measurement. The psychophysicist Stanley Smith Stevens defined nominal, ordinal, interval, and ratio scales. Nominal measurements do not have meaningful rank order among values, and permit any one-to-one transformation. Ordinal measurements have imprecise differences between consecutive values, but have a meaningful order to those values, and permit any order-preserving transformation. Interval measurements have meaningful distances between measurements defined, but the zero value is arbitrary (as in the case with longitude and temperature measurements in Celsius or Fahrenheit), and permit any linear transformation. Ratio measurements have both a meaningful zero value and the distances between different measurements defined, and permit any rescaling transformation.

Because variables conforming only to nominal or ordinal measurements cannot be reasonably measured numerically, sometimes they are grouped together as categorical variables, whereas ratio and interval measurements are grouped together as quantitative variables, which can be either discrete or continuous, due to their numerical nature. Such distinctions can often be loosely correlated with data type in computer science, in that dichotomous categorical variables may be represented with the Boolean data type, polytomous categorical variables with arbitrarily assigned integers in the integral data type, and continuous variables with the real data type involving floating point computation. But the mapping of computer science data types to statistical data types depends on which categorization of the latter is being implemented.

Other categorizations have been proposed. For example, Mosteller and Tukey (1977) distinguished grades, ranks, counted fractions, counts, amounts, and balances. Nelder (1990) described continuous counts, continuous ratios, count ratios, and categorical modes of data. See also Chrisman (1998), van den Berg (1991).

The issue of whether or not it is appropriate to apply different kinds of statistical methods to data obtained from different kinds of measurement procedures is complicated by issues concerning the transformation of variables and the precise interpretation of research questions. "The relationship between the data and what they describe merely reflects the fact that certain kinds of statistical statements may have truth values which are not invariant under some transformations. Whether or not a transformation is sensible to contemplate depends on the question one is trying to answer" (Hand, 2004, p. 82).

Simple Data Types

The following table classifies the various simple data types, associated distributions, permissible operations, etc. Regardless of the logical possible values, all of these data types are generally coded using real numbers, because the theory of random variables often explicitly assumes that they hold real numbers.

Data Type	Possible values	Example usage	Level of measurement	Distribution	Scale of relative differences	Permissible statistics	Regression analysis
binary	0, 1 (arbitrary labels)	binary outcome ("yes/no", "true/false", "success/failure", etc.)	nominal scale	Bernoulli	incomparable	mode, Chi-squared	logistic, probit
categorical	1, 2, ..., K (arbitrary labels)	categorical outcome (specific blood type, political party, word, etc.)		categorical			multinomial logit, multinomial probit
ordinal	integer or real number (arbitrary scale)	relative score, significant only for creating a ranking	ordinal scale	categorical??	relative comparison		ordinal regression (ordered logit, ordered probit)
binomial	0, 1, ..., N	number of successes (e.g. yes votes) out of *N* possible	interval scale??	binomial, beta-binomial, etc.	additive??	mean, median, mode, standard deviation, correlation	binomial regression (logistic, probit)
count	nonnegative integers (0, 1, ...)	number of items (telephone calls, people, molecules, births, deaths, etc.) in given interval/area/volume	ratio scale	Poisson, negative binomial, etc.	multiplicative	All statistics permitted for interval scales plus the following: geometric mean, harmonic mean, coefficient of variation	Poisson, negative binomial regression

real-valued additive	real number	temperature, relative distance, location parameter, etc. (or approximately, anything not varying over a large scale)	interval scale	normal, etc. (usually symmetric about the mean)	additive	mean, median, mode, standard deviation, correlation	standard linear regression
real-valued multiplicative	positive real number	price, income, size, scale parameter, etc. (especially when varying over a large scale)	ratio scale	log-normal, gamma, exponential, etc. (usually a skewed distribution)	multiplicative	All statistics permitted for interval scales plus the following: geometric mean, harmonic mean, coefficient of variation	generalized linear model with logarithmic link

Multivariate Data Types

Data that cannot be described using a single number are often shoehorned into random vectors of real-valued random variables, although there is an increasing tendency to treat them on their own. Some examples:

- Random vectors. The individual elements may or may not be correlated. Examples of distributions used to describe correlated random vectors are the multivariate normal distribution and multivariate t-distribution. In general, there may be arbitrary correlations between any elements and any others; however, this often becomes unmanageable above a certain size, requiring further restrictions on the correlated elements.
- Random matrices. Random matrices can be laid out linearly and treated as random vectors; however, this may not be an efficient way of representing the correlations between different elements. Some probability distributions are specifically designed for random matrices, e.g. the matrix normal distribution and Wishart distribution.
- Random sequences. These are sometimes considered to be the same as random vectors, but in other cases the term is applied specifically to cases where each random variable is only correlated with nearby variables (as in a Markov model). This is a particular case of a Bayes network and often used for very long sequences, e.g. gene sequences or lengthy text documents. A number of models are specifically designed for such sequences, e.g. hidden Markov models.
- Random processes. These are similar to random sequences, but where the length of the sequence is indefinite or infinite and the elements in the sequence are processed one-by-one. This is often used for data that can be described as a time series, e.g. the price of a stock on successive days. Random processes are also used to model values that vary continuously (e.g. the temperature at successive moments in time), rather than at discrete intervals.
- Bayes networks. These correspond to aggregates of random variables described using graphical models, where individual random variables are linked in a graph structure with conditional distributions relating variables to nearby variables.
- Multilevel models are subclasses of Bayes networks that can be thought of as having multiple levels of linear regression.
- Random trees. These are a subclass of Bayes network, where the variables are linked in a tree structure. An example is the problem of parsing a sentence, when statistical parsing techniques are used, such as probabilistic context-free grammars (PCFG's).
- Random fields. These represent the extension of random processes to multiple dimensions, and are common in physics, where they are used in statistical mechanics to describe properties such as force or electric field that can vary continuously over three dimensions (or four dimensions, when time is included).

These concepts originate in various scientific fields and frequently overlap in usage. As a result, it is very often the case that multiple concepts could potentially be applied to the same problem.

TERMINOLOGY AND THEORY OF INFERENTIAL STATISTICS

Statistics, Estimators and Pivotal Quantities

Consider independent identically distributed (IID) random variables with a given probability distribution: standard statistical inference and estimation theory defines a random sample as the random vector given by the column vector of these IID variables. The population being examined is described by a probability distribution that may have unknown parameters.

A statistic is a random variable that is a function of the random sample, but *not a function of unknown parameters*. The probability distribution of the statistic, though, may have unknown parameters.

Consider now a function of the unknown parameter: an estimator is a statistic used to estimate such function. Commonly used estimators include sample mean, unbiased sample variance and sample covariance.

A random variable that is a function of the random sample and of the unknown parameter, but whose probability distribution *does not depend on the unknown parameter* is called a pivotal quantity or pivot. Widely used pivots include the z-score, the chi square statistic and Student's t-value.

Between two estimators of a given parameter, the one with lower mean squared error is said to be more efficient. Furthermore, an estimator is said to be unbiased if its expected value is equal to the true value of the unknown parameter being estimated, and asymptotically unbiased if its expected value converges at the limit to the true value of such parameter.

Other desirable properties for estimators include: UMVUE estimators that have the lowest variance for all possible values of the parameter to be estimated (this is usually an easier property to verify than efficiency) and consistent estimators which converges in probability to the true value of such parameter.

This still leaves the question of how to obtain estimators in a given situation and carry the computation, several methods have been proposed: the method of moments, the maximum likelihood method, the least squares method and the more recent method of estimating equations.

Null Hypothesis and Alternative Hypothesis

Interpretation of statistical information can often involve the development of a null hypothesis which is usually (but not necessarily) that no relationship exists among variables or that no change occurred over time.

The best illustration for a novice is the predicament encountered by a criminal trial. The null hypothesis, H_0, asserts that the defendant is innocent, whereas the alternative hypothesis, H_1, asserts that the defendant is guilty. The indictment comes because of suspicion of the guilt. The H_0 (status quo) stands in opposition to H_1 and is maintained unless H_1 is supported by evidence "beyond a reasonable doubt". However, "failure to reject H_0" in this case does not imply innocence, but merely that the evidence was insufficient to convict. So the jury does not necessarily *accept* H_0 but *fails to reject* H_0. While one cannot "prove" a null hypothesis, one can test how close it is to being true with a power test, which tests for type II errors.

What statisticians call an alternative hypothesis is simply an hypothesis that contradicts the null hypothesis.

Error

Working from a null hypothesis, two basic forms of error are recognized:

- Type I errors where the null hypothesis is falsely rejected giving a "false positive".
- Type II errors where the null hypothesis fails to be rejected and an actual difference between populations is missed giving a "false negative".

Standard deviation refers to the extent to which individual observations in a sample differ from a central value, such as the sample or population mean, while Standard error refers to an estimate of difference between sample mean and population mean.

A statistical error is the amount by which an observation differs from its expected value, a residual is the amount an observation differs from the value the estimator of the expected value assumes on a given sample (also called prediction).

Mean squared error is used for obtaining efficient estimators, a widely used class of estimators. Root mean square error is simply the square root of mean squared error.

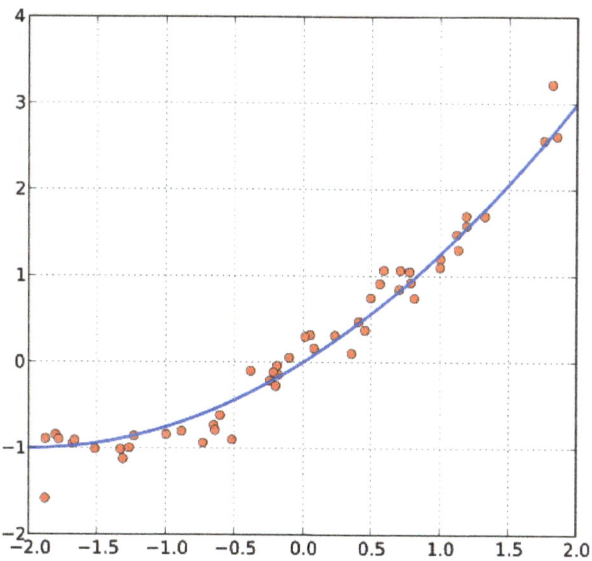

A least squares fit: in red the points to be fitted, in blue the fitted line.

Many statistical methods seek to minimize the residual sum of squares, and these are called "methods of least squares" in contrast to Least absolute deviations. The latter gives equal weight to small and big errors, while the former gives more weight to large errors. Residual sum of squares is also differentiable, which provides a handy property for doing regression. Least squares applied to linear regression is called ordinary least squares method and least squares applied to nonlinear regression is called non-linear least squares. Also in a linear regression model the non deterministic part of the model is called error term, disturbance or more simply noise. Both linear regression and non-linear regression are addressed in polynomial least squares, which also describes the variance in a prediction of the dependent variable (y axis) as a function of the independent variable (x axis) and the deviations (errors, noise, disturbances) from the estimated (fitted) curve.

Measurement processes that generate statistical data are also subject to error. Many of these errors are classified as random (noise) or systematic (bias), but other types of errors (e.g., blunder, such as when an analyst reports incorrect units) can also be important. The presence of missing data and/or censoring may result in biased estimates and specific techniques have been developed to address these problems.

Interval Estimation

Most studies only sample part of a population, so results don't fully represent the whole population. Any estimates obtained from the sample only approximate the population value. Confidence intervals allow statisticians to express how closely the sample estimate matches the true value in the whole population. Often they are expressed as 95% confidence intervals. Formally, a 95% confidence interval for a value is a range where, if the sampling and analysis were repeated under the same conditions (yielding a different dataset), the interval would include the true (population) value in 95% of all possible cases. This does *not* imply that the probability that the true value is in the confidence interval is 95%.

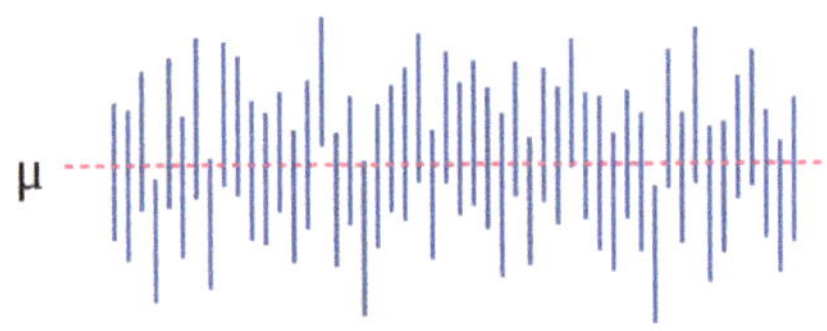

Confidence intervals: the red line is true value for the mean in this example, the blue lines are random confidence intervals for 100 realizations.

From the frequentist perspective, such a claim does not even make sense, as the true value is not a random variable. Either the true value is or is not within the given interval. However, it is true that, before any data are sampled and given a plan for how to construct the confidence interval, the probability is 95% that the yet-to-be-calculated interval will cover the true value: at this point, the limits of the interval are yet-to-be-observed random variables. One approach that does yield an interval that can be interpreted as having a given probability of containing the true value is to use a credible interval from Bayesian statistics: this approach depends on a different way of interpreting what is meant by "probability", that is as a Bayesian probability. In principle confidence intervals can be symmetrical or asymmetrical. An interval can be asymmetrical because it works as lower or upper bound for a parameter (left-sided interval or right sided interval), but it can also be asymmetrical because the two sided interval is built violating symmetry around the estimate. Sometimes the bounds for a confidence interval are reached asymptotically and these are used to approximate the true bounds.

Significance

Statistics rarely give a simple Yes/No type answer to the question under analysis. Interpretation often comes down to the level of statistical significance applied to the numbers and often refers to the probability of a value accurately rejecting the null hypothesis (sometimes referred to as the p-value).

> Important:
>
> **Pr (observation | hypothesis) ≠ Pr (hypothesis | observation)**
>
> The probability of observing a result given that some hypothesis is true is *not equivalent* to the probability that a hypothesis is true given that some result has been observed.
>
> Using the p-value as a "score" is committing an egregious logical error: **the transposed conditional fallacy.**

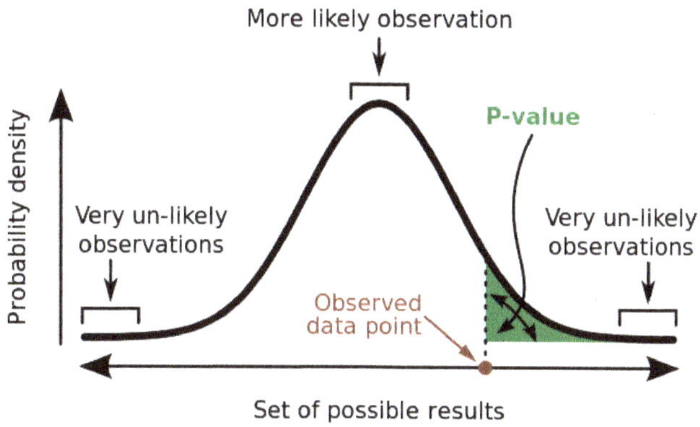

A **p-value** (shaded green area) is the probability of an observed (or more extreme) result assuming that the null hypothesis is true.

In this graph the black line is probability distribution for the test statistic, the critical region is the set of values to the right of the observed data point (observed value of the test statistic) and the p-value is represented by the green area.

The standard approach is to test a null hypothesis against an alternative hypothesis. A critical region is the set of values of the estimator that leads to refuting the null hypothesis. The probability of type I error is therefore the probability that the estimator belongs to the critical region given that null hypothesis is true (statistical significance) and the probability of type II error is the probability that the estimator doesn't belong to the critical region given that the alternative hypothesis is true. The statistical power of a test is the probability that it correctly rejects the null hypothesis when the null hypothesis is false. Referring to statistical significance does not necessarily mean that the overall result is significant in real world terms. For example, in a large study of a drug it may be shown that the drug has a statistically significant but very small beneficial effect, such that the drug is unlikely to help the patient noticeably.

While in principle the acceptable level of statistical significance may be subject to debate, the p-value is the smallest significance level that allows the test to reject the null hypothesis. This is logically equivalent to saying that the p-value is the probability, assuming the null hypothesis is true, of observing a result at least as extreme as the test statistic. Therefore, the smaller the p-value, the lower the probability of committing type I error.

Some problems are usually associated with this framework:

- A difference that is highly statistically significant can still be of no practical significance, but it is possible to properly formulate tests to account for this. One response involves going beyond reporting only the significance level to include the *p*-value when reporting whether a hypothesis is rejected or accepted. The p-value, however, does not indicate the size or importance of the observed effect and can also seem to exaggerate the importance of minor differences in large studies. A better and increasingly common approach is to report confidence intervals. Although these are produced from the same calculations as those of hypothesis tests or *p*-values, they describe both the size of the effect and the uncertainty surrounding it.
- Fallacy of the transposed conditional, aka prosecutor's fallacy: criticisms arise because the hypothesis testing approach forces one hypothesis (the null hypothesis) to be favored, since what is being evaluated is probability of the observed result given the null hypothesis and not probability of the null hypothesis given the observed result. An alternative to this approach is offered by Bayesian inference, although it requires establishing a prior probability.
- Rejecting the null hypothesis does not automatically prove the alternative hypothesis.
- As everything in inferential statistics it relies on sample size, and therefore under fat tails p-values may be seriously mis-computed.

Examples

Some well-known statistical tests and procedures are:

- Analysis of variance (ANOVA)
- Chi-squared test
- Correlation
- Factor analysis
- Mann–Whitney *U*
- Mean square weighted deviation (MSWD)
- Pearson product-moment correlation coefficient
- Regression analysis

- Spearman's rank correlation coefficient
- Student's *t*-test
- Time series analysis
- Conjoint Analysis

Misuse

Misuse of statistics can produce subtle, but serious errors in description and interpretation—subtle in the sense that even experienced professionals make such errors, and serious in the sense that they can lead to devastating decision errors. For instance, social policy, medical practice, and the reliability of structures like bridges all rely on the proper use of statistics.

Even when statistical techniques are correctly applied, the results can be difficult to interpret for those lacking expertise. The statistical significance of a trend in the data—which measures the extent to which a trend could be caused by random variation in the sample—may or may not agree with an intuitive sense of its significance. The set of basic statistical skills (and skepticism) that people need to deal with information in their everyday lives properly is referred to as statistical literacy.

There is a general perception that statistical knowledge is all-too-frequently intentionally misused by finding ways to interpret only the data that are favorable to the presenter. A mistrust and misunderstanding of statistics is associated with the quotation, "There are three kinds of lies: lies, damned lies, and statistics". Misuse of statistics can be both inadvertent and intentional, and the book *How to Lie with Statistics* outlines a range of considerations. In an attempt to shed light on the use and misuse of statistics, reviews of statistical techniques used in particular fields are conducted (e.g. Warne, Lazo, Ramos, and Ritter (2012)).

Ways to avoid misuse of statistics include using proper diagrams and avoiding bias. Misuse can occur when conclusions are overgeneralized and claimed to be representative of more than they really are, often by either deliberately or unconsciously overlooking sampling bias. Bar graphs are arguably the easiest diagrams to use and understand, and they can be made either by hand or with simple computer programs. Unfortunately, most people do not look for bias or errors, so they are not noticed. Thus, people may often believe that something is true even if it is not well represented. To make data gathered from statistics believable and accurate, the sample taken must be representative of the whole. According to Huff, "The dependability of a sample can be destroyed by [bias]... allow yourself some degree of skepticism."

To assist in the understanding of statistics Huff proposed a series of questions to be asked in each case:

- Who says so? (Does he/she have an axe to grind?)
- How does he/she know? (Does he/she have the resources to know the facts?)
- What's missing? (Does he/she give us a complete picture?)
- Did someone change the subject? (Does he/she offer us the right answer to the wrong problem?)

- Does it make sense? (Is his/her conclusion logical and consistent with what we already know?)

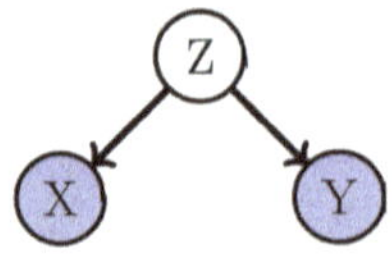

The confounding variable problem: X and Y may be correlated, not because there is causal relationship between them, but because both depend on a third variable Z. Z is called a confounding factor.

Misinterpretation: Correlation

The concept of correlation is particularly noteworthy for the potential confusion it can cause. Statistical analysis of a data set often reveals that two variables (properties) of the population under consideration tend to vary together, as if they were connected. For example, a study of annual income that also looks at age of death might find that poor people tend to have shorter lives than affluent people. The two variables are said to be correlated; however, they may or may not be the cause of one another. The correlation phenomena could be caused by a third, previously unconsidered phenomenon, called a lurking variable or confounding variable. For this reason, there is no way to immediately infer the existence of a causal relationship between the two variables.

HISTORY OF STATISTICAL SCIENCE

Statistical methods date back at least to the 5th century BC.

Some scholars pinpoint the origin of statistics to 1663, with the publication of *Natural and Political Observations upon the Bills of Mortality* by John Graunt. Early applications of statistical thinking revolved around the needs of states to base policy on demographic and economic data, hence its *stat-* etymology. The scope of the discipline of statistics broadened in the early 19th century to include the collection and analysis of data in general. Today, statistics is widely employed in government, business, and natural and social sciences.

Its mathematical foundations were laid in the 17th century with the development of the probability theory by Gerolamo Cardano, Blaise Pascal and Pierre de Fermat. Mathematical probability theory arose from the study of games of chance, although the concept of probability was already examined in medieval law and by philosophers such as Juan Caramuel.The method of least squares was first described by Adrien-Marie Legendre in 1805.

Gerolamo Cardano, the earliest pioneer on the mathematics of probability.

Karl Pearson, a founder of mathematical statistics.

The modern field of statistics emerged in the late 19th and early 20th century in three stages. The first wave, at the turn of the century, was led by the work of Francis Galton and Karl Pearson, who transformed statistics into a rigorous mathematical discipline used for analysis, not just in science, but in industry and politics as well. Galton's contributions included introducing the concepts of standard deviation, correlation, regression analysis and the application of these methods to the study of the variety of human characteristics – height, weight, eyelash length among others. Pearson developed the Pearson product-moment correlation coefficient, defined as a product-moment, the method of moments for the fitting of distributions to samples and the Pearson distribution, among many other things. Galton and Pearson founded *Biometrika* as the first journal of mathematical statistics and biostatistics (then called biometry), and the latter founded the world's first university statistics department at University College London. Ronald Fisher coined the term null hypothesis during the Lady tasting tea experiment, which "is never proved or established, but is possibly disproved, in the course of experimentation".

The second wave of the 1910s and 20s was initiated by William Gosset, and reached its culmination in the insights of Ronald Fisher, who wrote the textbooks that were to define the academic discipline in universities around the world. Fisher's most important publications were his 1918 seminal paper *The Correlation between Relatives on the Supposition of Mendelian Inheritance*, which was the first to use the statistical term, variance, his classic 1925 work *Statistical Methods for Research Workers* and his 1935 *The Design of Experiments*, where he developed rigorous design of experiments models. He originated the concepts of sufficiency, ancillary statistics, Fisher's linear discriminator and Fisher information. In his 1930 book *The Genetical Theory of Natural Selection* he applied statistics to various biological concepts such as Fisher's principle). Nevertheless, A. W. F. Edwards has remarked that it is "probably the most celebrated argument in evolutionary biology".(about the sex ratio), the Fisherian runaway, a concept in sexual selection about a positive feedback runaway affect found in evolution.

The final wave, which mainly saw the refinement and expansion of earlier developments, emerged from the collaborative work between Egon Pearson and Jerzy Neyman in the 1930s. They introduced the concepts of "Type II" error, power of a test and confidence intervals. Jerzy Neyman in 1934 showed that stratified random sampling was in general a better method of estimation than purposive (quota) sampling. Today, statistical methods are applied in all fields that involve decision making, for making accurate inferences from a collated body of data and for making decisions in the face of uncertainty based on statistical methodology. The use of modern computers has expedited large-scale statistical computations, and has also made possible new methods that are impractical to perform manually. Statistics continues to be an area of active research, for example on the problem of how to analyze Big data.

APPLICATIONS

Applied Statistics, Theoretical Statistics and Mathematical Statistics

"Applied statistics" comprises descriptive statistics and the application of inferential statistics. *Theoretical statistics* concerns both the logical arguments underlying justification of approaches to statistical inference, as well encompassing *mathematical statistics*. Mathematical statistics includes not only the manipulation of probability distributions necessary for deriving results related to methods of estimation and inference, but also various aspects of computational statistics and the design of experiments.

Machine Learning and Data Mining

There are two applications for machine learning and data mining: data management and data analysis. Statistics tools are necessary for the data analysis.

Statistics in Society

Statistics is applicable to a wide variety of academic disciplines, including natural and social sciences, government, and business. Statistical consultants can help organizations and companies that don't have in-house expertise relevant to their particular questions.

Statistical Computing

The rapid and sustained increases in computing power starting from the second half of the 20th century have had a substantial impact on the practice of statistical science. Early statistical models were almost always from the class of linear models, but powerful computers, coupled with suitable numerical algorithms, caused an increased interest in nonlinear models (such as neural networks) as well as the creation of new types, such as generalized linear models and multilevel models.

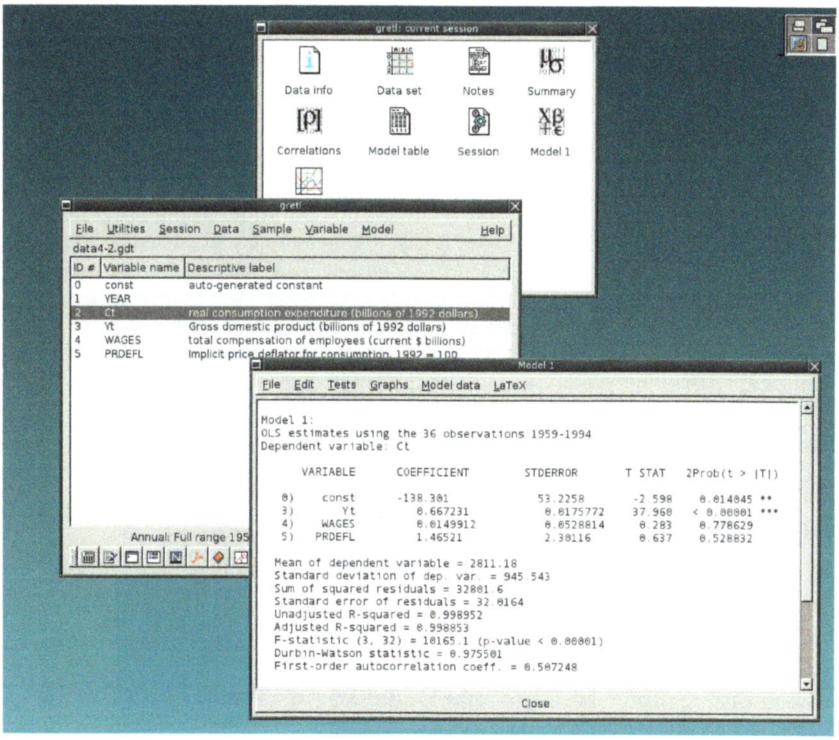

gretl, an example of an open source statistical package

Increased computing power has also led to the growing popularity of computationally intensive methods based on resampling, such as permutation tests and the bootstrap, while techniques such as Gibbs sampling have made use of Bayesian models more feasible. The computer revolution has implications for the future of statistics with new emphasis on "experimental" and "empirical" statistics. A large number of both general and special purpose statistical software are now available.

Statistics Applied to Mathematics or the Arts

Traditionally, statistics was concerned with drawing inferences using a semi-standardized methodology that was "required learning" in most sciences. This has changed with use of statistics in non-inferential contexts. What was once considered a dry subject, taken in many fields as a degree-requirement, is now viewed enthusiastically. Initially derided by some mathematical purists, it is now considered essential methodology in certain areas.

- In number theory, scatter plots of data generated by a distribution function may be transformed with familiar tools used in statistics to reveal underlying patterns, which may then lead to hypotheses.
- Methods of statistics including predictive methods in forecasting are combined with chaos theory and fractal geometry to create video works that are considered to have great beauty.
- The process art of Jackson Pollock relied on artistic experiments whereby underlying distributions in nature were artistically revealed. With the advent of computers, statistical methods were applied to formalize such distribution-driven natural processes to make and analyze moving video art.
- Methods of statistics may be used predicatively in performance art, as in a card trick based on a Markov process that only works some of the time, the occasion of which can be predicted using statistical methodology.
- Statistics can be used to predicatively create art, as in the statistical or stochastic music invented by Iannis Xenakis, where the music is performance-specific. Though this type of artistry does not always come out as expected, it does behave in ways that are predictable and tunable using statistics.

REFERENCES

1. Dodge, Y. (2006) The Oxford Dictionary of Statistical Terms, OUP. ISBN 0-19-920613-9
2. Romijn, Jan-Willem (2014). "Philosophy of statistics". Stanford Encyclopedia of Philosophy.
3. "Definition of STATISTICS". www.merriam-webster.com. Retrieved 2016-05-28.
4. "Essay on Statistics: Meaning and Definition of Statistics". Economics Discussion. 2014-12-02. Retrieved 2016-05-28.
5. Lund Research Ltd. "Descriptive and Inferential Statistics". statistics.laerd.com. Retrieved 2014-03-23.
6. "What Is the Difference Between Type I and Type II Hypothesis Testing Errors?". About.com Education. Retrieved 2015-11-27.
7. Moses, Lincoln E. (1986) Think and Explain with Statistics, Addison-Wesley, ISBN 978-0-201-15619-5 . pp. 1–3
8. Hays, William Lee, (1973) Statistics for the Social Sciences, Holt, Rinehart and Winston, p.xii, ISBN 978-0-03-077945-9

9. Moore, David (1992). "Teaching Statistics as a Respectable Subject". In F. Gordon and S. Gordon. Statistics for the Twenty-First Century. Washington, DC: The Mathematical Association of America. pp. 14–25. ISBN 978-0-88385-078-7.
10. Chance, Beth L.; Rossman, Allan J. (2005). "Preface". Investigating Statistical Concepts, Applications, and Methods (PDF). Duxbury Press. ISBN 978-0-495-05064-3.
11. Lakshmikantham,, ed. by D. Kannan,... V. (2002). Handbook of stochastic analysis and applications. New York: M. Dekker. ISBN 0824706609.
12. Schervish, Mark J. (1995). Theory of statistics (Corr. 2nd print. ed.). New York: Springer. ISBN 0387945466.
13. Freedman, D.A. (2005) Statistical Models: Theory and Practice, Cambridge University Press. ISBN 978-0-521-67105-7
14. McCarney R, Warner J, Iliffe S, van Haselen R, Griffin M, Fisher P (2007). "The Hawthorne Effect: a randomised, controlled trial". BMC Med Res Methodol. 7 (1): 30. doi:10.1186/1471-2288-7-30. PMC 1936999Freely accessible. PMID 17608932.
15. Rothman, Kenneth J; Greenland, Sander; Lash, Timothy, eds. (2008). "7". Modern Epidemiology (3rd ed.). Lippincott Williams & Wilkins. p. 100.
16. Mosteller, F., & Tukey, J. W. (1977). Data analysis and regression. Boston: Addison-Wesley.
17. Nelder, J. A. (1990). The knowledge needed to computerise the analysis and interpretation of statistical information. In Expert systems and artificial intelligence: the need for information about data. Library Association Report, London, March, 23–27.
18. Chrisman, Nicholas R (1998). "Rethinking Levels of Measurement for Cartography". Cartography and Geographic Information Science. 25 (4): 231–242. doi:10.1559/152304098782383043.
19. van den Berg, G. (1991). Choosing an analysis method. Leiden: DSWO Press
20. Hand, D. J. (2004). Measurement theory and practice: The world through quantification. London, UK: Arnold.
21. Piazza Elio, Probabilità e Statistica, Esculapio 2007
22. Everitt, Brian (1998). The Cambridge Dictionary of Statistics. Cambridge, UK New York: Cambridge University Press. ISBN 0521593468.
23. http://www.yourstatsguru.com/epar/rp-reviewed/cohen1994/
24. Rubin, Donald B.; Little, Roderick J. A.,Statistical analysis with missing data, New York: Wiley 2002
25. Ioannidis, J. P. A. (2005). "Why Most Published Research Findings Are False". PLoS Medicine. 2 (8): e124. doi:10.1371/journal.pmed.0020124. PMC 1182327Freely accessible. PMID 16060722.
26. Huff, Darrell (1954) How to Lie with Statistics, WW Norton & Company, Inc. New York, NY. ISBN 0-393-31072-8

27. Warne, R. Lazo; Ramos, T.; Ritter, N. (2012). “Statistical Methods Used in Gifted Education Journals, 2006–2010”. Gifted Child Quarterly. 56 (3): 134–149. doi:10.1177/0016986212444122.
28. Drennan, Robert D. (2008). “Statistics in archaeology”. In Pearsall, Deborah M. Encyclopedia of Archaeology. Elsevier Inc. pp. 2093–2100. ISBN 978-0-12-373962-9.
29. Cohen, Jerome B. (December 1938). “Misuse of Statistics”. Journal of the American Statistical Association. JSTOR. 33 (204): 657–674. doi:10.1080/01621459.1938.10502344.
30. Freund, J. E. (1988). “Modern Elementary Statistics”. Credo Reference.
31. Huff, Darrell; Irving Geis (1954). How to Lie with Statistics. New York: Norton. The dependability of a sample can be destroyed by [bias]... allow yourself some degree of skepticism.
32. Huff, Darrell; Irving Geis (1954). How to Lie with Statistics. New York: Norton.
33. Willcox, Walter (1938) “The Founder of Statistics”. Review of the International Statistical Institute 5(4):321–328. JSTOR 1400906
34. J. Franklin, The Science of Conjecture: Evidence and Probability before Pascal,Johns Hopkins Univ Pr 2002
35. Helen Mary Walker (1975). Studies in the history of statistical method. Arno Press.
36. Galton, F (1877). “Typical laws of heredity”. Nature. 15: 492–553. doi:10.1038/015492a0.
37. Stigler, S. M. (1989). “Francis Galton’s Account of the Invention of Correlation”. Statistical Science. 4 (2): 73–79. doi:10.1214/ss/1177012580.
38. Pearson, K. (1900). “On the Criterion that a given System of Deviations from the Probable in the Case of a Correlated System of Variables is such that it can be reasonably supposed to have arisen from Random Sampling”. Philosophical Magazine Series 5. 50 (302): 157–175. doi:10.1080/14786440009463897.
39. “Karl Pearson (1857–1936)”. Department of Statistical Science – University College London.
40. Fisher|1971|loc=Chapter II. The Principles of Experimentation, Illustrated by a Psycho-physical Experiment, Section 8. The Null Hypothesis
41. OED quote: 1935 R. A. Fisher, The Design of Experiments ii. 19, “We may speak of this hypothesis as the ‘null hypothesis’, and it should be noted that the null hypothesis is never proved or established, but is possibly disproved, in the course of experimentation.”
42. Stanley, J. C. (1966). “The Influence of Fisher’s “The Design of Experiments” on Educational Research Thirty Years Later”. American Educational Research Journal. 3 (3): 223. doi:10.3102/00028312003003223.
43. Box, JF (February 1980). “R. A. Fisher and the Design of Experiments, 1922-1926”. The American Statistician. 34 (1): 1–7. doi:10.2307/2682986. JSTOR

2682986.

44. Yates, F (June 1964). "Sir Ronald Fisher and the Design of Experiments". Biometrics. 20 (2): 307–321. doi:10.2307/2528399. JSTOR 2528399.
45. Stanley, Julian C. (1966). "The Influence of Fisher's "The Design of Experiments" on Educational Research Thirty Years Later". American Educational Research Journal. 3 (3): 223–229. doi:10.3102/00028312003003223. JSTOR 1161806.
46. Agresti, Alan; David B. Hichcock (2005). "Bayesian Inference for Categorical Data Analysis" (PDF). Statistical Methods & Applications. 14 (14): 298. doi:10.1007/s10260-005-0121-y.
47. Edwards, A.W.F. (1998). "Natural Selection and the Sex Ratio: Fisher's Sources". American Naturalist. 151 (6): 564–569. doi:10.1086/286141. PMID 18811377.
48. Fisher, R.A. (1915) The evolution of sexual preference. Eugenics Review (7) 184:192
49. Fisher, R.A. (1930) The Genetical Theory of Natural Selection. ISBN 0-19-850440-3
50. Edwards, A.W.F. (2000) Perspectives: Anecdotal, Historial and Critical Commentaries on Genetics. The Genetics Society of America (154) 1419:1426
51. Andersson, M. (1994) Sexual selection. ISBN 0-691-00057-3
52. Andersson, M. and Simmons, L.W. (2006) Sexual selection and mate choice. Trends, Ecology and Evolution (21) 296:302
53. Gayon, J. (2010) Sexual selection: Another Darwinian process. Comptes Rendus Biologies (333) 134:144
54. Neyman, J (1934). "On the two different aspects of the representative method: The method of stratified sampling and the method of purposive selection". Journal of the Royal Statistical Society. 97 (4): 557–625. JSTOR 2342192.
55. "Science in a Complex World - Big Data: Opportunity or Threat?". Santa Fe Institute.
56. Nikoletseas, M. M. (2014) "Statistics: Concepts and Examples." ISBN 978-1500815684
57. Anderson, D.R.; Sweeney, D.J.; Williams, T.A. (1994) Introduction to Statistics: Concepts and Applications, pp. 5–9. West Group. ISBN 978-0-314-03309-3

## CHAPTER 2

Descriptive Statistics

Chapter Contents

- **Introduction**
- **Use in Statistical Analysis**
- **Continuous Data**

Descriptive statistics are brief descriptive coefficients that summarize a given data set, which can be either a representation of the entire population or a sample of it. Descriptive statistics are broken down into measures of central tendency and measures of variability, or spread. Measures of central tendency include the mean, median and mode, while measures of variability include the standard deviation or variance, the minimum and maximum variables, and the kurtosis and skewness.

INTRODUCTION

Descriptive statistics are statistics that quantitatively describe or summarize features of a collection of information. Descriptive statistics are distinguished from inferential statistics (or inductive statistics), in that descriptive statistics aim to summarize a sample, rather than use the data to learn about the population that the sample of data is thought to represent. This generally means that descriptive statistics, unlike inferential statistics, are not developed on the basis of probability theory. Even when a data analysis draws its main conclusions using inferential statistics, descriptive statistics are generally also presented. For example in papers reporting on human subjects, typically a table is included giving the overall sample size, sample sizes in important subgroups (e.g., for each treatment or exposure group), and demographic or clinical characteristics such as the average age, the proportion of subjects of each sex, the proportion of subjects with related comorbidities etc.

Some measures that are commonly used to describe a data set are measures of central tendency and measures of variability or dispersion. Measures of central tendency include the mean, median and mode, while measures of variability include the standard deviation (or variance), the minimum and maximum values of the variables, kurtosis and skewness.

USE IN STATISTICAL ANALYSIS

Descriptive statistics provide simple summaries about the sample and about the observations that have been made. Such summaries may be either quantitative, i.e. summary statistics, or visual, i.e. simple-to-understand graphs. These summaries may either form the basis of the initial description of the data as part of a more extensive statistical analysis, or they may be sufficient in and of themselves for a particular investigation.

For example, the shooting percentage in basketball is a descriptive statistic that summarizes the performance of a player or a team. This number is the number of shots made divided by the number of shots taken. For example, a player who shoots 33% is making approximately one shot in every three. The percentage summarizes or describes multiple discrete events. Consider also the grade point average. This single number describes the general performance of a student across the range of their course experiences.

The use of descriptive and summary statistics has an extensive history and, indeed, the simple tabulation of populations and of economic data was the first way the topic of statistics appeared. More recently, a collection of summarization techniques has been formulated under the heading of exploratory data analysis: an example of such a technique is the box plot. In the business world, descriptive statistics provides a useful summary of many types of data. For example, investors and brokers may use a historical account of return behavior by performing empirical and analytical analyses on their investments in order to make better investing decisions in the future.

Univariate Analysis

Univariate analysis involves describing the distribution of a single variable, including its central tendency (including the mean, median, and mode) and dispersion (including the range and quantiles of the data-set, and measures of spread such as the variance and standard deviation). The shape of the distribution may also be described via indices such as skewness and kurtosis. Characteristics of a variable's distribution may also be depicted in graphical or tabular format, including histograms and stem-and-leaf display.

Bivariate and Multivariate Analysis

When a sample consists of more than one variable, descriptive statistics may be used to describe the relationship between pairs of variables. In this case, descriptive statistics include:

- Cross-tabulations and contingency tables
- Graphical representation via scatterplots
- Quantitative measures of dependence
- Descriptions of conditional distributions

The main reason for differentiating univariate and bivariate analysis is that bivariate analysis is not only simple descriptive analysis, but also it describes the relationship between two different variables. Quantitative measures of dependence include correlation (such as Pearson's r when both variables are continuous, or Spearman's rho if one or both are not) and covariance (which reflects the scale variables are measured on). The slope, in regression analysis, also reflects the relationship between variables. The unstandardized slope indicates the unit change in the criterion variable for a one unit change in the predictor. The standardized slope indicates this change in standardized (z-score) units. Highly skewed data are often transformed by taking logarithms. Use of logarithms makes graphs more symmetrical and look more similar to the normal distribution, making them easier to interpret intuitively.

CONTINUOUS DATA

Central Tendency

In statistics, a central tendency (or, more commonly, a measure of central tendency) is a central or typical value for a probability distribution. It may also be called a center or location of the distribution. Colloquially, measures of central tendency are often called averages. The term central tendency dates from the late 1920s.

The most common measures of central tendency are the arithmetic mean, the median and the mode. A central tendency can be calculated for either a finite set of values or for a theoretical distribution, such as the normal distribution. Occasionally authors use central tendency to denote "the tendency of quantitative data to cluster around some central value."

The central tendency of a distribution is typically contrasted with its dispersion or variability; dispersion and central tendency are the often characterized properties of distributions. Analysts may judge whether data has a strong or a weak central tendency based on its dispersion.

Measures of Central Tendency

The following may be applied to one-dimensional data. Depending on the circumstances, it may be appropriate to transform the data before calculating a central tendency. Examples are squaring the values or taking logarithms. Whether a transformation is appropriate and what it should be, depend heavily on the data being analyzed.

- Arithmetic mean (or simply, mean) – the sum of all measurements divided by the number of observations in the data set.

- Median – the middle value that separates the higher half from the lower half of the data set. The median and the mode are the only measures of central tendency that can be used for ordinal data, in which values are ranked relative to each other but are not measured absolutely.
- Mode – the most frequent value in the data set. This is the only central tendency measure that can be used with nominal data, which have purely qualitative category assignments.
- Geometric mean – the nth root of the product of the data values, where there are n of these. This measure is valid only for data that are measured absolutely on a strictly positive scale.
- Harmonic mean – the reciprocal of the arithmetic mean of the reciprocals of the data values. This measure too is valid only for data that are measured absolutely on a strictly positive scale.
- Weighted arithmetic mean – an arithmetic mean that incorporates weighting to certain data elements.
- Truncated mean (or trimmed mean) – the arithmetic mean of data values after a certain number or proportion of the highest and lowest data values have been discarded.
    - Interquartile mean – a truncated mean based on data within the interquartile range.
- Midrange – the arithmetic mean of the maximum and minimum values of a data set.
- Midhinge – the arithmetic mean of the two quartiles.
- Trimean – the weighted arithmetic mean of the median and two quartiles.
- Winsorized mean – an arithmetic mean in which extreme values are replaced by values closer to the median.

Any of the above may be applied to each dimension of multi-dimensional data, but the results may not be invariant to rotations of the multi-dimensional space. In addition, there is the

- Geometric median - which minimizes the sum of distances to the data points. This is the same as the median when applied to one-dimensional data, but it is not the same as taking the median of each dimension independently. It is not invariant to different rescaling of the different dimensions.

The Quadratic mean (often known as the root mean square) is useful in engineering, but is not often used in statistics. This is because it is not a good indicator of the center of the distribution when the distribution includes negative values.

Solutions to Variational Problems

Several measures of central tendency can be characterized as solving a variational problem, in the sense of the calculus of variations, namely minimizing variation from the center. That is, given a measure of statistical dispersion, one asks for a measure of central tendency that minimizes variation: such that variation from the center is minimal among all choices of center. In a quip, "dispersion precedes location". In the sense of L^p spaces, the correspondence is:

L^p	dispersion	central tendency
L^1	average absolute deviation	median
L^2	standard deviation	mean
L^∞	maximum deviation	midrange

Thus standard deviation about the mean is lower than standard deviation about any other point, and the maximum deviation about the midrange is lower than the maximum deviation about any other point. The uniqueness of this characterization of mean follows from convex optimization. Indeed, for a given (fixed) data set x, the function

$$f_2(c) = \|x - c\|_2$$

represents the dispersion about a constant value c relative to the L^2 norm. Because the function f_2 is a strictly convex coercive function, the minimizer exists and is unique.

Note that the median in this sense is not in general unique, and in fact any point between the two central points of a discrete distribution minimizes average absolute deviation. The dispersion in the L^1 norm, given by

$$f_1(c) = \|x - c\|_1$$

is not strictly convex, whereas strict convexity is needed to ensure uniqueness of the minimizer. In spite of this, the minimizer is unique for the L^∞ norm.

Mean

In mathematics, mean has several different definitions depending on the context.

In probability and statistics, mean and expected value are used synonymously to refer to one measure of the central tendency either of a probability distribution or of the random variable characterized by that distribution. In the case of a discrete probability distribution of a random variable X, the mean is equal to the sum over every possible value weighted by the probability of that value; that is, it is computed by taking the product of each possible value x of X and its probability P(x), and then adding all these products together, giving $\mu = \sum xP(x)$. An analogous formula applies to the case of a continuous probability distribution. Not every probability distribution has a defined mean; see the Cauchy distribution for an example. Moreover, for some distributions the mean is infinite: for example, when the probability of the value 2^n is $\frac{1}{2^n}$ for n = 1, 2, 3, ...

For a data set, the terms arithmetic mean, mathematical expectation, and sometimes average are used synonymously to refer to a central value of a discrete set of numbers: specifically, the sum of the values divided by the number of values. The arithmetic mean of a set of numbers x_1, x_2, ..., x_n is typically denoted by $\bar{x}$, pronounced "x bar". If the data set were based on a series of observations obtained by sampling from a statistical population, the arithmetic mean is termed the sample mean (denoted $\bar{x}$) to distinguish it from the population mean (denoted μ or μ_x).

For a finite population, the population mean of a property is equal to the arithmetic mean of the given property while considering every member of the population. For example, the population mean height is equal to the sum of the heights of every individual divided by the total number of individuals. The sample mean may differ from the population mean, especially for small samples. The law of large numbers dictates that the larger the size of the sample, the more likely it is that the sample mean will be close to the population mean.

Outside of probability and statistics, a wide range of other notions of "mean" are often used in geometry and analysis; examples are given below.

Types of Mean

Arithmetic Mean (AM)

The arithmetic mean (or simply "mean") of a sample $x_1, x_2, \ldots, x_n$, usually denoted by $\bar{x}$, is the sum of the sampled values divided by the number of items in the sample:

$$\bar{x} = \frac{x_1 + x_2 + \cdots + x_n}{n}$$

For example, the arithmetic mean of five values: 4, 36, 45, 50, 75 is

$$\frac{4 + 36 + 45 + 50 + 75}{5} = \frac{210}{5} = 42.$$

Geometric Mean (GM)

The geometric mean is an average that is useful for sets of positive numbers that are interpreted according to their product and not their sum (as is the case with the arithmetic mean) e.g. rates of growth.

$$\bar{x} = \left(\prod_{i=1}^{n} x_i\right)^{\frac{1}{n}}$$

For example, the geometric mean of five values: 4, 36, 45, 50, 75 is:

$$(4 \times 36 \times 45 \times 50 \times 75)^{1/5} = \sqrt[5]{24\,300\,000} = 30.$$

Harmonic Mean (HM)

The harmonic mean is an average which is useful for sets of numbers which are defined in relation to some unit, for example speed (distance per unit of time).

$$\bar{x} = n \cdot \left(\sum_{i=1}^{n} \frac{1}{x_i} \right)^{-1}$$

For example, the harmonic mean of the five values: 4, 36, 45, 50, 75 is

$$\frac{5}{\frac{1}{4} + \frac{1}{36} + \frac{1}{45} + \frac{1}{50} + \frac{1}{75}} = \frac{5}{\frac{1}{3}} = 15.$$

Relationship between AM, GM, and HM

AM, GM, and HM satisfy these inequalities:

$$AM \geq GM \geq HM$$

Equality holds if and only if all the elements of the given sample are equal.

Statistical Location

In descriptive statistics, the mean may be confused with the median, mode or mid-range, as any of these may be called an "average" (more formally, a measure of central tendency). The mean of a set of observations is the arithmetic average of the values; however, for skewed distributions, the mean is not necessarily the same as the middle value (median), or the most likely value (mode).

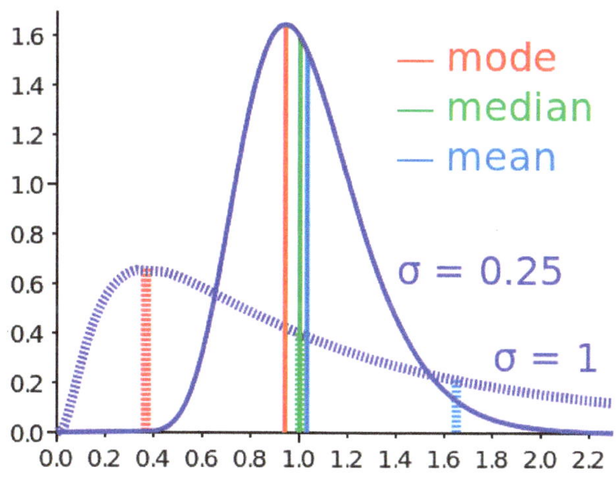

Comparison of the arithmetic mean, median and mode of two skewed (log-normal) distributions.

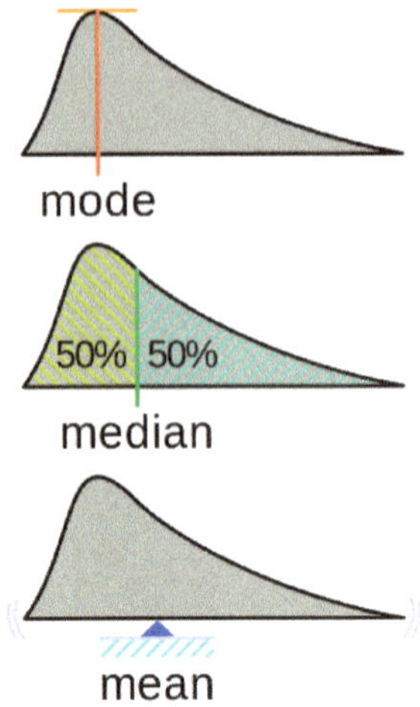

Geometric visualization of the mode, median and mean of an arbitrary probability density function.

For example, mean income is typically skewed upwards by a small number of people with very large incomes, so that the majority have an income lower than the mean. By contrast, the median income is the level at which half the population is below and half is above. The mode income is the most likely income, and favors the larger number of people with lower incomes. While the median and mode are often more intuitive measures for such skewed data, many skewed distributions are in fact best described by their mean, including the exponential and Poisson distributions.

Mean of a Probability Distribution

The mean of a probability distribution is the long-run arithmetic average value of a random variable having that distribution. In this context, it is also known as the expected value. For a discrete probability distribution, the mean is given by $\sum xP(x)$, where the sum is taken over all possible values of the random variable and P(x) is the probability mass function. For a continuous distribution, the mean is $\int_{-\infty}^{\infty} xf(x)\,dx$, where f'(x) is the probability density function. In all cases, including those in which the distribution is neither discrete nor continuous, the mean is the Lebesgue integral of the random variable with respect to its probability measure. The mean need not exist or be finite; for some probability distributions the mean is infinite (+∞ or −∞), while others have no mean.

Generalized Means

Power Mean: The generalized mean, also known as the power mean or Hölder mean, is an abstraction of the quadratic, arithmetic, geometric and harmonic means. It is defined for a set of n positive numbers x_i by

$$\bar{x}(m) = \left(\frac{1}{n} \cdot \sum_{i=1}^{n} x_i^m\right)^{\frac{1}{m}}$$

By choosing different values for the parameter m, the following types of means are obtained:

$m \to \infty$ maximum of x_i

$m = 2$ quadratic mean

$m = 1$ arithmetic mean

$m \to 0$ geometric mean

$m = -1$ harmonic mean

$m \to -\infty$ minimum of x_i

f-Mean

This can be generalized further as the generalized f-mean

$$\bar{x} = f^{-1}\left(\frac{1}{n} \cdot \sum_{i=1}^{n} f(x_i)\right)$$

and again a suitable choice of an invertible *f* will give

$f(x) = x$ arithmetic mean,

$f(x) = \frac{1}{x}$ harmonic mean,

$f(x) = x^m$ power mean,

$f(x) = \ln x$ geometric mean.

Weighted Arithmetic Mean

The weighted arithmetic mean (or weighted average) is used if one wants to combine average values from samples of the same population with different sample sizes:

$$\bar{x} = \frac{\sum_{i=1}^{n} w_i \cdot x_i}{\sum_{i=1}^{n} w_i}.$$

The weights w_i represent the sizes of the different samples. In other applications they represent a measure for the reliability of the influence upon the mean by the respective values.

Truncated Mean

Sometimes a set of numbers might contain outliers, i.e., data values which are much lower or much higher than the others. Often, outliers are erroneous data caused by artifacts. In this case, one can use a truncated mean. It involves discarding given parts of the data at the top or the bottom end, typically an equal amount at each end, and then taking the arithmetic mean of the remaining data. The number of values removed is indicated as a percentage of total number of values.

Interquartile Mean

The interquartile mean is a specific example of a truncated mean. It is simply the arithmetic mean after removing the lowest and the highest quarter of values.

$$\bar{x} = \frac{2}{n} \sum_{i=(n/4)+1}^{3n/4} x_i$$

assuming the values have been ordered, so is simply a specific example of a weighted mean for a specific set of weights.

Mean of a Function

In some circumstances mathematicians may calculate a mean of an infinite (even an uncountable) set of values. This can happen when calculating the mean value y_{ave} of a function f(x). Intuitively this can be thought of as calculating the area under a section of a curve and then dividing by the length of that section. This can be done crudely by counting squares on graph paper or more precisely by integration. The integration formula is written as:

$$y_{\text{ave}}(a,b) = \frac{\int_a^b f(x)\,dx}{b-a}$$

Care must be taken to make sure that the integral converges. But the mean may be finite even if the function itself tends to infinity at some points.

Mean of Angles and Cyclical Quantities

Angles, times of day, and other cyclical quantities require modular arithmetic to add and otherwise combine numbers. In all these situations, there will not be a unique mean. For example, the times an hour before and after midnight are equidistant to both midnight and noon. It is also possible that no mean exists. Consider a color wheel -- there is no mean to the set of all colors. In these situations, you must decide which mean is most useful. You can do this by adjusting the values before averaging, or by using a specialized approach for the mean of circular quantities.

Fréchet Mean

The Fréchet mean gives a manner for determining the "center" of a mass distribution on a surface or, more generally, Riemannian manifold. Unlike many other means, the Fréchet mean is defined on a space whose elements cannot necessarily be added together or multiplied by scalars. It is sometimes also known as the Karcher mean (named after Hermann Karcher).

Distribution of the Sample Mean

The arithmetic mean of a population, or population mean, is denoted μ. The sample mean (the arithmetic mean of a sample of values drawn from the population) makes a good estimator of the population mean, as its expected value is equal to the population mean (that is, it is an unbiased estimator). The sample mean is a random variable, not a constant, since its calculated value will randomly differ depending on which members of the population are sampled, and consequently it will have its own distribution. For a random sample of n observations from a normally distributed population, the sample mean distribution is normally distributed with mean and variance as follows:

$$\bar{x} \sim N\left\{\mu, \frac{\sigma^2}{n}\right\}.$$

Often, since the population variance is an unknown parameter, it is estimated by the mean sum of squares; when this estimated value is used, the distribution of the sample mean is no longer a normal distribution but rather a Student's t distribution with n − 1 degrees of freedom.

Median

The median is the value separating the higher half of a data sample, a population, or a probability distribution, from the lower half. In simple terms, it may be thought of as the "middle" value of a data set. For example, in the data set {1, 3, 3, 6, 7, 8, 9}, the median is 6, the fourth number in the sample. The median is a commonly used measure of the properties of a data set in statistics and probability theory.

The basic advantage of the median in describing data compared to the mean (often simply described as the "average") is that it is not skewed so much by extremely large or small values, and so it may give a better idea of a 'typical' value. For example, in understanding statistics like household income or assets which vary greatly, a mean may be skewed by a small number of extremely high or low values. Median income, for example, may be a better way to suggest what a 'typical' income is.

1, 3, 3, **6**, 7, 8, 9

Median = 6

1, 2, 3, **4**, **5**, 6, 8, 9

Median = (4 + 5) ÷ 2

= 4.5

Finding the median in sets of data with an odd and even number of values.

Because of this, the median is of central importance in robust statistics, as it is the most resistant statistic, having a breakdown point of 50%: so long as no more than half the data are contaminated, the median will not give an arbitrarily large or small result.

Basic Procedure

The median of a finite list of numbers can be found by arranging all the numbers from smallest to greatest.

If there is an odd number of numbers, the middle one is picked. For example, consider the set of numbers:

1, 3, 3, 6, 7, 8, 9

This set contains seven numbers. The median is the fourth of them, which is 6.

If there are an even number of observations, then there is no single middle value; the median is then usually defined to be the mean of the two middle values. For example, in the data set:

1, 2, 3, 4, 5, 6, 8, 9

The median is the mean of the middle two numbers: this is (4 + 5) ÷ 2, which is 4.5. (In more technical terms, this interprets the median as the fully trimmed mid-range.)

The formula used to find the middle number of a data set of n numbers is (n + 1) ÷ 2. This either gives the middle number (for an odd number of values) or the halfway point between the two middle values. For example, with 14 values, the formula will give 7.5, and the median will be taken by averaging the seventh and eighth values.

You will also be able to find the median using the Stem-and-Leaf Plot.

There is no widely accepted standard notation for the median, but some authors represent the median of a variable x either as $\tilde{x}$ or as $\mu_{1/2}$ sometimes also M. In any of these cases, the use of these or other symbols for the median needs to be explicitly defined when they are introduced.

Discussion

The median is used primarily for skewed distributions, which it summarizes differently from the arithmetic mean. Consider the multiset { 1, 2, 2, 2, 3, 14 }. The median is 2 in this case, (as is the mode), and it might be seen as a better indication of central tendency (less susceptible to the exceptionally large value in data) than the arithmetic mean of 4.

Calculation of medians is a popular technique in summary statistics and summarizing statistical data, since it is simple to understand and easy to calculate, while also giving a measure that is more robust in the presence of outlier values than is the mean. The widely cited empirical relationship between the relative locations of the mean and the median for skewed distributions is, however, not generally true. There are, however, various relationships for the absolute difference between them; see below.

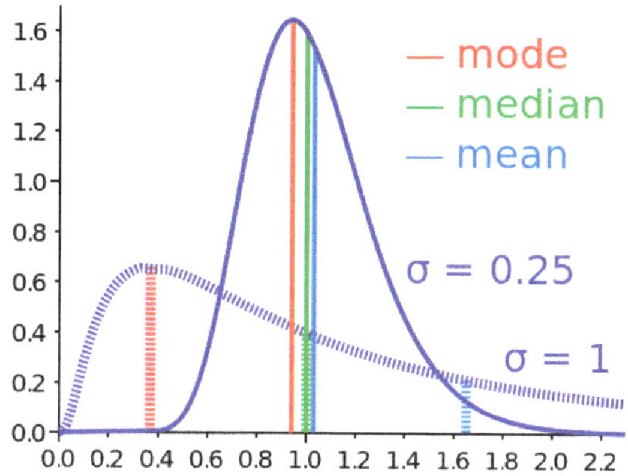

Comparison of mean, median and mode of two log-normal distributions with different skewness.

The median does not identify a specific value within the data set, since more than one value can be at the median level and with an even number of observations (as shown above) no value need be exactly at the value of the median. Nonetheless, the value of the median is uniquely determined with the usual definition. A related concept, in which the outcome is forced to correspond to a member of the sample, is the medoid.

In a population, at most half have values strictly less than the median and at most half have values strictly greater than it. If each group contains less than half the population, then some of the population is exactly equal to the median. For example, if a < b < c, then the median of the list {a, b, c} is b, and, if a < b < c < d, then the median of the list {a, b, c, d} is the mean of b and c; i.e., it is (b + c)/2. Indeed, as it is based on the middle data in a group, it is not necessary to even know the value of extreme results in order to calculate a median. For example, in a psychology test investigating the time needed to solve a problem, if a small number of people failed to solve the problem at all in the given time a median can still be calculated.

The median can be used as a measure of location when a distribution is skewed, when end-values are not known, or when one requires reduced importance to be attached to outliers, e.g., because they may be measurement errors.

A median is only defined on ordered one-dimensional data, and is independent of any distance metric. A geometric median, on the other hand, is defined in any number of dimensions. The median is one of a number of ways of summarizing the typical values associated with members of a statistical population; thus, it is a possible location parameter. The median is the 2nd quartile, 5th decile, and 50th percentile. Since the median is the same as the second quartile, its calculation is illustrated in the quartiles. A median can be worked out for ranked but not numerical classes (e.g. working out a median grade when students are graded from A to F), although the result might be halfway between grades if there are an even number of cases.

When the median is used as a location parameter in descriptive statistics, there are several choices for a measure of variability: the range, the interquartile range, the mean absolute deviation, and the median absolute deviation.

For practical purposes, different measures of location and dispersion are often compared on the basis of how well the corresponding population values can be estimated from a sample of data. The median, estimated using the sample median, has good properties in this regard. While it is not usually optimal if a given population distribution is assumed, its properties are always reasonably good. For example, a comparison of the efficiency of candidate estimators shows that the sample mean is more statistically efficient than the sample median when data are uncontaminated by data from heavy-tailed distributions or from mixtures of distributions, but less efficient otherwise, and that the efficiency of the sample median is higher than that for a wide range of distributions. More specifically, the median has a 64% efficiency compared to the minimum-variance mean (for large normal samples), which is to say the variance of the median will be ~50% greater than the variance of the mean.

Probability Distributions

For any probability distribution on the real line **R** with cumulative distribution function F, regardless of whether it is any kind of continuous probability distribution, in particular an absolutely continuous distribution (which has a probability density function), or a discrete probability distribution, a median is by definition any real number m that satisfies the inequalities

$$\mathrm{P}(X \leq m) \geq \frac{1}{2} \text{ and } \mathrm{P}(X \geq m) \geq \frac{1}{2}$$

or, equivalently, the inequalities

$$\int_{(-\infty,m]} dF(x) \geq \frac{1}{2} \text{ and } \int_{[m,\infty)} dF(x) \geq \frac{1}{2}$$

in which a Lebesgue–Stieltjes integral is used. For an absolutely continuous probability distribution with probability density function f, the median satisfies

$$\mathrm{P}(X \leq m) = \mathrm{P}(X \geq m) = \int_{-\infty}^{m} f(x)\, dx = \frac{1}{2}.$$

Any probability distribution on **R** has at least one median, but there may be more than one median. Where exactly one median exists, statisticians speak of "the median" correctly; even when the median is not unique, some statisticians speak of "the median" informally.

Medians of Particular Distributions

The medians of certain types of distributions can be easily calculated from their parameters; furthermore, they exist even for some distributions lacking a well-defined mean, such as the Cauchy distribution:

- The median of a symmetric distribution which possesses a mean μ also takes the value μ.
    - The median of a normal distribution with mean μ and variance σ^2 is μ. In fact, for a normal distribution, mean = median = mode.
    - The median of a uniform distribution in the interval [a, b] is (a + b) / 2, which is also the mean.
- The median of a Cauchy distribution with location parameter x_0 and scale parameter y is x_0, the location parameter.
- The median of a power law distribution x^{-a}, with exponent a > 1 is $2^{1/(a-1)}x_{min}$, where x_{min} is the minimum value for which the power law holds
- The median of an exponential distribution with rate parameter λ is the natural logarithm of 2 divided by the rate parameter: $\lambda^{-1}\ln 2$.
- The median of a Weibull distribution with shape parameter k and scale parameter λ is $\lambda(\ln 2)^{1/k}$.

Populations

Optimality Property: The mean absolute error of a real variable c with respect to the random variable X is

$$E(|X-c|)$$

Provided that the probability distribution of X is such that the above expectation exists, then m is a median of X if and only if m is a minimizer of the mean absolute error with respect to X. In particular, m is a sample median if and only if m minimizes the arithmetic mean of the absolute deviations.

More generally, a median is defined as a minimum of

$$E(|X-c|-|X|),$$

as discussed below in the section on multivariate medians (specifically, the spatial median).

This optimization-based definition of the median is useful in statistical data-analysis, for example, in k-medians clustering.

Unimodal Distributions: It can be shown for a unimodal distribution that the median $\tilde{X}$ and the mean $\bar{X}$ lie within $(3/5)^{1/2} \approx 0.7746$ standard deviations of each other. In symbols,

$$\frac{|\tilde{X} - \bar{X}|}{\sigma} \leq (3/5)^{1/2}$$

where |.| is the absolute value.

A similar relation holds between the median and the mode: they lie within $3^{1/2} \approx 1.732$ standard deviations of each other:

$$\frac{|\tilde{X} - \text{mode}|}{\sigma} \leq 3^{1/2}.$$

Inequality Relating Means and Medians

If the distribution has finite variance, then the distance between the median and the mean is bounded by one standard deviation.

This bound was proved by Mallows, who used Jensen's inequality twice, as follows. We have

$$\begin{aligned} |\mu - m| = |\mathrm{E}(X - m)| &\leq \mathrm{E}(|X - m|) \\ &\leq \mathrm{E}(|X - \mu|) \\ &\leq \sqrt{\mathrm{E}((X - \mu)^2)} = \sigma. \end{aligned}$$

The first and third inequalities come from Jensen's inequality applied to the absolute-value function and the square function, which are each convex. The second inequality comes from the fact that a median minimizes the absolute deviation function

$$a \mapsto \mathrm{E}(|X - a|).$$

This proof can easily be generalized to obtain a multivariate version of the inequality, as follows:

$$\begin{aligned} \|\mu - m\| = \|\mathrm{E}(X - m)\| &\leq \mathrm{E}\,\|X - m\| \\ &\leq \mathrm{E}(\|X - \mu\|) \\ &\leq \sqrt{\mathrm{E}(\|X - \mu\|^2)} = \sqrt{\mathrm{trace}(\mathrm{var}(X))} \end{aligned}$$

where m is a spatial median, that is, a minimizer of the function $a \mapsto \mathrm{E}(\|X - a\|)$ The spatial median is unique when the data-set's dimension is two or more. An alternative proof uses the one-sided Chebyshev inequality; it appears in an inequality on location and scale parameters.

Jensen's Inequality for Medians

Jensen's inequality states that for any random variable x with a finite expectation E(x) and for any convex function f

$$f(E(x)) \leq E(f(x))$$

It has been shown that if x is a real variable with a unique median m and f is a C function then

$$f(m) \leq \text{Median}(f(x))$$

A C function is a real valued function, defined on the set of real numbers R, with the property that for any real t

$$f^{-1}((-\infty, t]) = \{x \in R \mid f(x) \leq t\}$$

is a closed interval, a singleton or an empty set.

Medians for Samples

Efficient Computation of the Sample Median

Even though comparison-sorting n items requires Ω(n log n) operations, selection algorithms can compute the k'th-smallest of n items with only Θ(n) operations. This includes the median, which is the n/2'th order statistic (or for an even number of samples, the average of the two middle order statistics).

Selection algorithms still have the downside of requiring Ω(n) memory, that is, they need to have the full sample (or a linear-sized portion of it) in memory. Because this, as well as the linear time requirement, can be prohibitive, several estimation procedures for the median have been developed. A simple one is the median of three rule, which estimates the median as the median of a three-element subsample; this is commonly used as a subroutine in the quicksort sorting algorithm, which uses an estimate of its input's median. A more robust estimator is Tukey's ninther, which is the median of three rule applied with limited recursion: if A is the sample laid out as an array, and

med3(A) = median(A, A[n/2], A[n]),

then

ninther(A) = med3(med3(A[1 ... 1/3n]), med3(A[1/3n ... 2/3n]), med3(A[2/3n ... n]))

The remedian is an estimator for the median that requires linear time but sub-linear memory, operating in a single pass over the sample.

Easy Explanation of the Sample Median

In individual series (if number of observation is very low) first one must arrange all the observations in order. Then count(n) is the total number of observation in given data.

If n is odd then Median (M) = value of ((n + 1)/2)th item term.

If n is even then Median (M) = value of [((n)/2)th item term + ((n)/2 + 1)th item term]/2

For an odd number of values

As an example, we will calculate the sample median for the following set of observations: 1, 5, 2, 8, 7.

Start by sorting the values: 1, 2, 5, 7, 8.

In this case, the median is 5 since it is the middle observation in the ordered list.

The median is the ((n + 1)/2)th item, where n is the number of values. For example, for the list {1, 2, 5, 7, 8}, we have n = 5, so the median is the ((5 + 1)/2)th item.

median = (6/2)th item

median = 3rd item

median = 5

For an even number of values

As an example, we will calculate the sample median for the following set of observations: 1, 6, 2, 8, 7, 2.

Start by sorting the values: 1, 2, 2, 6, 7, 8.

In this case, the arithmetic mean of the two middlemost terms is (2 + 6)/2 = 4. Therefore, the median is 4 since it is the arithmetic mean of the middle observations in the ordered list.

We also use this formula MEDIAN = {(n + 1)/2}th item . n = number of values

As above example 1, 2, 2, 6, 7, 8 n = 6 Median = {(6 + 1)/2}th item = 3.5th item. In this case, the median is average of the 3rd number and the next one (the fourth number). The median is (2 + 6)/2 which is 4.

Sampling Distribution

The distribution of both the sample mean and the sample median were determined by Laplace. The distribution of the sample median from a population with a density function f(x) is asymptotically normal with mean m and variance

$$\frac{1}{4nf(m)^2}$$

where m is the median value of distribution and n is the sample size. In practice $f(m) = \frac{1}{2}$ by definition

These results have also been extended. It is now known for the p th quantile that the distribution of the sample p-th quantile is asymptotically normal around the p-th quantile with variance equal to

$$\frac{p(1-p)}{nf(x_p)^2}$$

where $f(x_p)$ is the value of the distribution density at the p-th quantile.

In the case of a discrete variable, the sampling distribution of the median for small-samples can be investigated as follows. We take the sample size to be an odd number N = 2n+1. If a given value v is to be the median of the sample then two conditions must be satisfied. The first is that at most n observations can have a value of v-1 or less.

The second is that at most n observations can have a value of v+1 or more. Let i be the number of observations which have a value of v-1 or less and let k be the number of observations which have a value of v+1 or more. Then i and k both have a minimum value of 0 and a maximum of n. If an observation has a value below v, it is not relevant how far below v it is and conversely, if an observation has a value above v, it is not relevant how far above v it is. We can therefore represent the observations as following a trinomial distribution with probabilities $F(v-1)$, $f(v)$ and $1-F(v)$. The probability that the median m will have a value v is then given by

$$\Pr(m=v)=\sum_{i=0}^{n}\sum_{k=0}^{n}\frac{N!}{i!(N-i-k)!k!}[F(v-1)]^{i}[f(v)]^{N-i-k}[1-F(v)]^{k}.$$

Summing this over all values of v defines a proper distribution and gives a unit sum. In practice, the function f(v) will often not be known but it can be estimated from an observed frequency distribution. An example is given in the following table where the actual distribution is not known but a sample of 3,800 observations allows a sufficiently accurate assessment of f(v).

v	0	0.5	1	1.5	2	2.5	3	3.5	4	4.5	5
f(v)	0.000	0.008	0.010	0.013	0.083	0.108	0.328	0.220	0.202	0.023	0.005
F(v)	0.000	0.008	0.018	0.031	0.114	0.222	0.550	0.770	0.972	0.995	1.000

Using these data it is possible to investigate the effect of sample size on the standard errors of the mean and median. The observed mean is 3.16, the observed raw median is 3 and the observed interpolated median is 3.174. The following table gives some comparison statistics. The standard error of the median is given both from the above expression for $pr(m=v)$ and from the asymptotic approximation given earlier.

Sample size **Statistic**	**3**	9	**15**	**21**
Expected value of median	3.198	3.191	3.174	3.161
Standard error of median (above formula)	0.482	0.305	0.257	0.239
Standard error of median (asymptotic approximation)	0.879	0.508	0.393	0.332
Standard error of mean	0.421	0.243	0.188	0.159

The expected value of the median falls slightly as sample size increases while, as would be expected, the standard errors of both the median and the mean are proportionate to the inverse square root of the sample size. The asymptotic approximation errs on the side of caution by overestimating the standard error.

In the case of a continuous variable, the following argument can be used. If a given value v is to be the median, then one observation must take the value v. The elemental probability of this is f(v)dv. Then, of the remaining 2n observations, exactly n of them must be above v and the remaining n below. The probability of this is the nth term of a binomial distribution with parameters F(v) and 2n. Finally we multiply by 2n+1 since any of the observations in the sample can be the median observation. Hence the elemental probability of the median at the point v is given by

$$f(v)\frac{(2n)!}{n!n!}[F(v)]^n[1-F(v)]^n(2n+1)\,dv.$$

Now we introduce the beta function. For integer arguments α and β, this can be expressed as $\mathrm{B}(\alpha,\beta)=(\alpha-1)!(\beta-1)!/(\alpha+\beta-1)!$. Also, we note that $f(v)=dF(v)/dv$. Using these relationships and setting both α and β equal to $(n+1)$ allows the last expression to be written as

$$\frac{[F(v)]^n[1-F(v)]^n}{\mathrm{B}(n+1,n+1)}\,dF(v)$$

Hence the density function of the median is a symmetric beta distribution over the unit interval which supports F(v). Its mean, as we would expect, is 0.5 and its standard deviation (which is the standard error of the sample median) is $1/(2\sqrt{N+2})$. However this finding can only be used if (i) f(v) is known or can be assumed (ii) f(v) can be integrated to find F(v) and (iii) F(v) can in turn be inverted. This will not always be the case and even when it is, the cut-off points for f(v) can be calculated directly without recourse to the distribution of the median on the unit interval. So while interesting in theory, this result is not much use in practice.

Estimation of Variance from Sample Data

The value of $(2f(x))^{-2}$—the asymptotic value of $n^{-\frac{1}{2}}(\nu-m)$ where ν is the population median—has been studied by several authors. The standard 'delete one' jackknife method produces inconsistent results. An alternative—the 'delete k' method—where k grows with the sample size has been shown to be asymptotically consistent. This method may be computationally expensive for large data sets. A bootstrap estimate is known to be consistent, but converges very slowly (order of $n^{-\frac{1}{4}}$). Other methods have been proposed but their behavior may differ between large and small samples.

Efficiency

The efficiency of the sample median, measured as the ratio of the variance of the mean to the variance of the median, depends on the sample size and on the underlying population distribution. For a sample of size $N=2n+1$ from the normal distribution, the ratio is

$$\frac{4n}{\pi(2n+1)}$$

For large samples (as n tends to infinity) this ratio tends to $\frac{2}{\pi}$

Other Estimators

For univariate distributions that are symmetric about one median, the Hodges–Lehmann estimator is a robust and highly efficient estimator of the population median.

If data are represented by a statistical model specifying a particular family of probability distributions, then estimates of the median can be obtained by fitting that family of probability distributions to the data and calculating the theoretical median of the fitted distribution. Pareto interpolation is an application of this when the population is assumed to have a Pareto distribution.

Coefficient of Dispersion

The coefficient of dispersion (CD) is defined as the ratio of the average absolute deviation from the median to the median of the data. It is a statistical measure used by the states of Iowa, New York and South Dakota in estimating dues taxes. In symbols

$$CD = \frac{1}{n}\frac{\sum |m-x|}{m}$$

where n is the sample size, m is the sample median and x is a variate. The sum is taken over the whole sample.

Confidence intervals for a two sample test where the sample sizes are large have been derived by Bonett and Seier This test assumes that both samples have the same median but differ in the dispersion around it. The confidence interval (CI) is bounded inferiorly by

$$\exp\left[\log\left(\frac{t_a}{t_b}\right) - z_\alpha\left(\operatorname{var}\left[\log\left(\frac{t_a}{t_b}\right)\right]\right)^{0.5}\right]$$

where t_j is the mean absolute deviation of the jth sample, var() is the variance and z_α is the value from the normal distribution for the chosen value of α: for $\alpha = 0.05$, $z_\alpha = 1.96$. The following formulae are used in the derivation of these confidence intervals

$$\operatorname{var}[\log(t_a)] = \frac{\left(\frac{s_a^2}{t_a^2} + \left(\frac{x_a - \bar{x}}{t_a}\right)^2 - 1\right)}{n}$$

$$\operatorname{var}\left[\log\left(\frac{t_a}{t_b}\right)\right] = \operatorname{var}[\log(t_a)] + \operatorname{var}[\log(t_b)] - 2r(\operatorname{var}[\log(t_a)]\operatorname{var}[\log(t_b)])^{0.5}$$

where r is the Pearson correlation coefficient between the squared deviation scores

$$d_{ia} = |x_{ia} - \bar{x}_a| \text{ and } d_{ib} = |x_{ib} - \bar{x}_b|$$

a and b here are constants equal to 1 and 2, x is a variate and s is the standard deviation of the sample.

Multivariate median

Previously, this discussed the univariate median, when the sample or population had one-dimension. When the dimension is two or higher, there are multiple concepts that extend the definition of the univariate median; each such multivariate median agrees with the univariate median when the dimension is exactly one.

Marginal Median

The marginal median is defined for vectors defined with respect to a fixed set of coordinates. A marginal median is defined to be the vector whose components are univariate medians. The marginal median is easy to compute, and its properties were studied by Puri and Sen.

Spatial Median

For N vectors in a normed vector space, a spatial median minimizes the average distance

$$a \mapsto \frac{1}{N} \sum_n (\|x_n - a\|),$$

where x_n and a are vectors. The spatial median is unique when the data-set's dimension is two or more and the norm is the Euclidean norm (or another strictly convex norm). The spatial median is also called the L1 median, even when the norm is Euclidean. Other names are used especially for finite sets of points: geometric median, Fermat point (in mechanics), or Weber or Fermat-Weber point (in geographical location theory).

More generally, a spatial median is defined as a minimizer of

$$a \mapsto \frac{1}{N} \sum_n (\|x_n - a\| - \|x_n\|);$$

this general definition is convenient for defining a spatial median of a population in a finite-dimensional normed space, for example, for distributions without a finite mean. Spatial medians are defined for random vectors with values in a Banach space.

The spatial median is a robust and highly efficient estimator of a central tendency of a population.

Other Multivariate Medians

An alternative generalization of the spatial median in higher dimensions that does not relate to a particular metric is the centerpoint.

Interpolated Median

When dealing with a discrete variable, it is sometimes useful to regard the observed values as being midpoints of underlying continuous intervals. An example of this is a Likert scale where opinions or preferences are expressed on a scale with a set number of possible responses. If the scale consists of the positive integers, an observation of 3 might be regarded as representing the interval from 2.50 to 3.50. It is possible to estimate the median of the underlying variable. If say 22% of the observations are of value 2 or below and 55.0% are of 3 or below (so 33% have the value 3), then the median m is 3 since the median is the smallest value of x for which F(x) is greater than a half. But the interpolated median is somewhere between 2.50 and 3.50. First we add half of the interval width w to the median to get the upper bound of the median interval. Then we subtract that proportion of the interval width which equals the proportion of the 33% which lies above the 50% mark. In other words, we split up the interval width pro rata to the numbers of observations. In this case, the 33% is split into 28% below the median and 5% above it so we subtract 5/33 of the interval width from the upper bound of 3.50 to give an interpolated median of 3.35. More formally, if the values f(x) are known, the interpolated median can be calculated from

$$m_{\text{int}} = m + w\left[\frac{1}{2} - \frac{F(m) - 1/2}{f(m)}\right].$$

Alternatively, if in an observed sample there are k scores above the median category, j scores in it and i scores below it then the interpolated median is given by

$$m_{\text{int}} = m + \frac{w}{2}\left[\frac{k - i}{j}\right].$$

Pseudo-Median

For univariate distributions that are symmetric about one median, the Hodges–Lehmann estimator is a robust and highly efficient estimator of the population median; for non-symmetric distributions, the Hodges–Lehmann estimator is a robust and highly efficient estimator of the population pseudo-median, which is the median of a symmetrized distribution and which is close to the population median. The Hodges–Lehmann estimator has been generalized to multivariate distributions.

Variants of Regression

The Theil–Sen estimator is a method for robust linear regression based on finding medians of slopes.

Median Filter

In the context of image processing of monochrome raster images there is a type of noise, known as the salt and pepper noise, when each pixel independently becomes black (with some small probability) or white (with some small probability), and is unchanged otherwise (with the probability close to 1). An image constructed of median values of neighborhoods (like 3×3 square) can effectively reduce noise in this case.

Cluster Analysis

In cluster analysis, the k-medians clustering algorithm provides a way of defining clusters, in which the criterion of maximizing the distance between cluster-means that is used in k-means clustering, is replaced by maximizing the distance between cluster-medians.

Median–Median Line

This is a method of robust regression. The idea dates back to Wald in 1940 who suggested dividing a set of bivariate data into two halves depending on the value of the independent parameter x: a left half with values less than the median and a right half with values greater than the median. He suggested taking the means of the dependent y and independent x variables of the left and the right halves and estimating the slope of the line joining these two points. The line could then be adjusted to fit the majority of the points in the data set.

Nair and Shrivastava in 1942 suggested a similar idea but instead advocated dividing the sample into three equal parts before calculating the means of the subsamples. Brown and Mood in 1951 proposed the idea of using the medians of two subsamples rather the means. Tukey combined these ideas and recommended dividing the sample into three equal size subsamples and estimating the line based on the medians of the subsamples.

Median-Unbiased Estimators

Any mean-unbiased estimator minimizes the risk (expected loss) with respect to the squared-error loss function, as observed by Gauss. A median-unbiased estimator minimizes the risk with respect to the absolute-deviation loss function, as observed by Laplace. Other loss functions are used in statistical theory, particularly in robust statistics.

The theory of median-unbiased estimators was revived by George W. Brown in 1947:

An estimate of a one-dimensional parameter θ will be said to be median-unbiased if, for fixed θ, the median of the distribution of the estimate is at the value θ; i.e., the estimate underestimates just as often as it overestimates. This requirement seems for most purposes to accomplish as much as the mean-unbiased requirement and has the additional property that it is invariant under one-to-one transformation.

Further properties of median-unbiased estimators have been reported. Median-unbiased estimators are invariant under one-to-one transformations.

There are methods of construction median-unbiased estimators that are optimal (in a sense analogous to minimum-variance property considered for mean-unbiased estimators). Such constructions exist for probability distributions having monotone likelihood-functions. One such procedure is an analogue of the Rao–Blackwell procedure for mean-unbiased estimators: The procedure holds for a smaller class of probability distributions than does the Rao—Blackwell procedure but for a larger class of loss functions.

Mode

The mode is the value that appears most often in a set of data. The mode of a discrete probability distribution is the value x at which its probability mass function takes its maximum value. In other words, it is the value that is most likely to be sampled. The mode of a continuous probability distribution is the value x at which its probability density function has its maximum value, so the mode is at the peak.

Like the statistical mean and median, the mode is a way of expressing, in a (usually) single number, important information about a random variable or a population. The numerical value of the mode is the same as that of the mean and median in a normal distribution, and it may be very different in highly skewed distributions.

The mode is not necessarily unique to a given distribution, since the probability mass function or probability density function may take the same maximum value at several points x_1, x_2, etc. The most extreme case occurs in uniform distributions, where all values occur equally frequently. When a probability density function has multiple local maxima it is common to refer to all of the local maxima as modes of the distribution. Such a continuous distribution is called multimodal (as opposed to unimodal).

In symmetric unimodal distributions, such as the normal distribution, the mean (if defined), median and mode all coincide. For samples, if it is known that they are drawn from a symmetric distribution, the sample mean can be used as an estimate of the population mode.

Mode of a Sample

The mode of a sample is the element that occurs most often in the collection. For example, the mode of the sample [1, 3, 6, 6, 6, 6, 7, 7, 12, 12, 17] is 6. Given the list of data [1, 1, 2, 4, 4] the mode is not unique - the dataset may be said to be bimodal, while a set with more than two modes may be described as multimodal.

For a sample from a continuous distribution, such as [0.935..., 1.211..., 2.430..., 3.668..., 3.874...], the concept is unusable in its raw form, since no two values will be exactly the same, so each value will occur precisely once. In order to estimate the mode, the usual practice is to discretize the data by assigning frequency values to intervals of equal distance, as for making a histogram, effectively replacing the values by the midpoints of the intervals they are assigned to. The mode is then the value where the histogram reaches its peak. For small or middle-sized samples the outcome of this procedure is sensitive to the choice of interval width if chosen too narrow or too wide; typically one should have a sizable fraction of the data concentrated in a relatively small number of intervals (5 to 10), while the fraction of the data falling outside these intervals is also sizable. An alternate approach is kernel density estimation, which essentially blurs point samples to produce a continuous estimate of the probability density function which can provide an estimate of the mode.

The following MATLAB (or Octave) code example computes the mode of a sample:

```
X = sort(x);
indices   =  find(diff([X; realmax]) > 0); % indices where repeated values
change
[modeL,i] =  max (diff([0; indices]));     % longest persistence length of
repeated values
mode      =  X(indices(i));
```

The algorithm requires as a first step to sort the sample in ascending order. It then computes the discrete derivative of the sorted list, and finds the indices where this derivative is positive. Next it computes the discrete derivative of this set of indices, locating the maximum of this derivative of indices, and finally evaluates the sorted sample at the point where that maximum occurs, which corresponds to the last member of the stretch of repeated values.

Comparison of Mean, Median and Mode

Comparison of common averages of values { 1, 2, 2, 3, 4, 7, 9 }			
Type	**Description**	**Example**	**Result**
Arithmetic mean	Sum of values of a data set divided by number of values: $\bar{x}=\frac{1}{n}\sum_{i=1}^{n} x_i$	(1+2+2+3+4+7+9) / 7	**4**
Median	Middle value separating the greater and lesser halves of a data set	1, 2, 2, **3**, 4, 7, 9	**3**
Mode	Most frequent value in a data set	1, **2**, **2**, 3, 4, 7, 9	**2**

Use

Unlike mean and median, the concept of mode also makes sense for "nominal data" (i.e., not consisting of numerical values in the case of mean, or even of ordered values in the case of median). For example, taking a sample of Korean family names, one might find that "Kim" occurs more often than any other name. Then "Kim" would be the mode of the sample. In any voting system where a plurality determines victory, a single modal value determines the victor, while a multi-modal outcome would require some tie-breaking procedure to take place.

Unlike median, the concept of mode makes sense for any random variable assuming values from a vector space, including the real numbers (a one-dimensional vector space) and the integers (which can be considered embedded in the reals). For example, a distribution of points in the plane will typically have a mean and a mode, but the concept of median does not apply. The median makes sense when there is a linear order on the possible values. Generalizations of the concept of median to higher-dimensional spaces are the geometric median and the centerpoint.

Uniqueness and Definedness

For some probability distributions, the expected value may be infinite or undefined, but if defined, it is unique. The mean of a (finite) sample is always defined. The median is the value such that the fractions not exceeding it and not falling below it are each at least 1/2. It is not necessarily unique, but never infinite or totally undefined. For a data sample it is the "halfway" value when the list of values is ordered in increasing value, where usually for a list of even length the numerical average is taken of the two values closest to "halfway". Finally, as said before, the mode is not necessarily unique. Certain pathological distributions (for example, the Cantor distribution) have no defined mode at all. For a finite data sample, the mode is one (or more) of the values in the sample.

Properties

Assuming definedness, and for simplicity uniqueness, the following are some of the most interesting properties.

- All three measures have the following property: If the random variable (or each value from the sample) is subjected to the linear or affine transformation which replaces X by aX+b, so are the mean, median and mode.
- However, if there is an arbitrary monotonic transformation, only the mode follows; for example, if X is replaced by exp(X), the mode changes from m to exp(m) but the mean and median won't.
- Except for extremely small samples, the mode is insensitive to "outliers" (such as occasional, rare, false experimental readings). The median is also very robust in the presence of outliers, while the mean is rather sensitive.
- In continuous unimodal distributions the median lies, as a rule of thumb, between the mean and the mode, about one third of the way going from mean to mode. In a formula, median ≈ (2 × mean + mode)/3. This rule, due to Karl Pearson, often applies to slightly non-symmetric distributions that resemble a normal distribution, but it is not always true and in general the three statistics can appear in any order.
- For unimodal distributions, the mode is within $\sqrt{3}$ standard deviations of the mean, and the root mean square deviation about the mode is between the standard deviation and twice the standard deviation.

Example for a Skewed Distribution

An example of a skewed distribution is personal wealth: Few people are very rich, but among those some are extremely rich. However, many are rather poor.

A well-known class of distributions that can be arbitrarily skewed is given by the log-normal distribution. It is obtained by transforming a random variable X having a normal distribution into random variable Y = e^X. Then the logarithm of random variable Y is normally distributed, hence the name.

Taking the mean μ of X to be 0, the median of Y will be 1, independent of the standard deviation σ of X. This is so because X has a symmetric distribution, so its median is also 0. The transformation from X to Y is monotonic, and so we find the median e^0 = 1 for Y.

When X has standard deviation σ = 0.25, the distribution of Y is weakly skewed. Using formulas for the log-normal distribution, we find:

$$\begin{aligned} \text{mean} &= e^{\mu+\sigma^2/2} &&= e^{0+0.25^2/2} &&\approx 1.032 \\ \text{mode} &= e^{\mu-\sigma^2} &&= e^{0-0.25^2} &&\approx 0.939 \\ \text{median} &= e^{\mu} &&= e^{0} &&= 1 \end{aligned}$$

Indeed, the median is about one third on the way from mean to mode.

When X has a larger standard deviation, $\sigma = 1$, the distribution of Y is strongly skewed. Now

$$\begin{aligned} \text{mean} &= e^{\mu+\sigma^2/2} &&= e^{0+1^2/2} &&\approx 1.649 \\ \text{mode} &= e^{\mu-\sigma^2} &&= e^{0-1^2} &&\approx 0.368 \\ \text{median} &= e^{\mu} &&= e^{0} &&= 1 \end{aligned}$$

Here, Pearson's rule of thumb fails.

Van Zwet Condition

Van Zwet derived an inequality which provides sufficient conditions for this inequality to hold. The inequality

Mode $\leq$ Median $\leq$ Mean

holds if

F(Median - x) + F(Median + x) ≥ 1

for all x where F() is the cumulative distribution function of the distribution.

Unimodal Distributions

It can be shown for a unimodal distribution that the median $\tilde{X}$ and the mean $\bar{X}$ lie within $(3/5)^{1/2} \approx 0.7746$ standard deviations of each other. In symbols,

$$\frac{|\tilde{X} - \bar{X}|}{\sigma} \leq (3/5)^{1/2}$$

where $|\cdot|$ is the absolute value.

A similar relation holds between the median and the mode: they lie within $3^{1/2} \approx 1.732$ standard deviations of each other:

$$\frac{|\tilde{X} - \text{mode}|}{\sigma} \leq 3^{1/2}.$$

Confidence Interval for the Mode with a Single Data Point

It is a common but false belief that from a single observation x we cannot gain information about the variability in the population and that consequently that finite length confidence intervals for mean and/or variance are impossible even in principle.

It is possible for an unknown unimodal distribution to estimate a confidence interval for the mode with a sample size of 1. This was first shown by Abbot and Rosenblatt and extended by Blachman and Machol. This confidence interval can be sharpened if the distribution can be assumed to be symmetrical. It is further possible to sharpen this interval if the distribution is normally distributed.

Let the confidence interval be $1 - \alpha$. Then the confidence intervals for the general, symmetric and normally distributed variates respectively are

$$X \pm \left(\frac{2}{\alpha} - 1\right)|X - \theta|$$

$$X \pm \left(\frac{1}{\alpha} - 1\right)|X - \theta|$$

$$X \pm \left(\frac{0.484}{\alpha} - 1\right)|X - \theta|$$

where X is the variate, θ is the mode and $|\cdot|$ is the absolute value.

These estimates are conservative. The confidence intervals for the mode at the 90% level given by these estimators are X ± 19 | X - θ |, X ± 9 | X - θ | and X ± 5.84 | X - θ | for the general, symmetric and normally distributed variates respectively. The 95% confidence interval for a normally distributed variate is given by X ± 10.7 | X - θ |. It may be worth noting that the mean and the mode coincide if the variates are normally distributed.

The 95% bound for a normally distributed variate has been improved and is now known to be X ± 9.68 | X - θ | The bound for a 99% confidence interval is X ± 48.39 | X - θ'|

Note

Machol has shown that given a known density symmetrical about 0 that given a single sample value (x) that the 90% confidence intervals of population mean are

$$x \pm 5|x - \nu|$$

where v is the population median.

If the precise form of the distribution is not known but it is known to be symmetrical about zero then we have

$$P(X - k|X - a| \leq \mu \leq X + k|X - a|) \geq 1 - \frac{1}{1+k}$$

where X is the variate, μ is the population mean and a and k are arbitrary real numbers.

It is also possible to estimate a confidence interval for the standard deviation from a single observation if the distribution is symmetrical about 0. For a normal distribution the with an unknown variance and a single data point (X) the 90%, 95% and 99% confidence intervals for the standard deviation are [0, 8|X|], [0, 17|X|] and [0, 70|X|]. These intervals may be shorted if the mean is known to be bounded by a multiple of the standard deviation.

If the distribution is known to be normal then it is possible to estimate a confidence interval for the mean and variance from a simple value. The 90% confidence intervals are

$$X - 23.3|X| \leq \mu \leq X + 23.3|X|$$

$$\sigma \leq 10|X|$$

The confidence intervals can be estimated for any chosen range.

This method is not limited to the normal distribution but can be used with any known distribution.

Statistical Tests

These estimators have been used to create hypothesis tests for simple samples from normal or symmetrical unimodal distributions. Let the distribution have an assumed mean (μ_0). The null hypothesis is that the assumed mean of the distribution lies within the confidence interval of the sample mean (m). The null hypothesis is accepted if

$$\mu_0 < \frac{x+m}{2} \pm k|x-m|$$

where x is the value of the sample and k is a constant. The null hypothesis is rejected if

$$\mu_0 > \frac{x+m}{2} \pm k|x-m|$$

The value of k depends on the choice of confidence interval and the nature of the assumed distribution.

If the distribution is assumed or is known to be normal then the values of k for the 50%, 66.6%, 75%, 80%, 90%, 95% and 99% confidence intervals are 0.50, 1.26, 1.80, 2.31, 4.79, 9.66 and 48.39 respectively.

If the distribution is assumed or known to be unimodal and symmetrical but not normal then the values of k for the 50%, 66.6%, 75%, 80%, 90%, 95% and 99% confidence intervals are 0.50, 1.87, 2.91, 3.94, 8.97, 18.99, 99.00 respectively.

To see how this test works we assume or know a priori that the population from which the sample is drawn has a mean of μ_0 and that the population has a symmetrical unimodal distribution - a class that includes the normal distribution. We wish to know if the mean estimated from the sample is representative of the population at a pre chosen level of confidence.

Assume that the distribution is normal and let the confidence interval be 95%. Then k = 9.66.

Assuming that the sample is representative of the population, the sample mean (m) will then lie within the range determined from the formula:

$$\mu_0 < \frac{x+m}{2} \pm 9.66|x-m|$$

If subsequent sampling shows that the sample mean lies outside these parameters the sample mean is to be considered to differ significantly from the population mean.

Statistical Dispersion

In statistics, dispersion (also called variability, scatter, or spread) denotes how stretched or squeezed a distribution (theoretical or that underlying a statistical sample) is. Common examples of measures of statistical dispersion are the variance, standard deviation and interquartile range.

Dispersion is contrasted with location or central tendency, and together they are the most used properties of distributions.

Measures of Statistical Dispersion

A measure of statistical dispersion is a nonnegative real number that is zero if all the data are the same and increases as the data become more diverse.

Most measures of dispersion have the same units as the quantity being measured. In other words, if the measurements are in meters or seconds, so is the measure of dispersion. Such measures of dispersion include:

- Sample standard deviation
- Interquartile range (IQR)
- Range
- Mean absolute difference (also known as Gini mean absolute difference)
- Median absolute deviation (MAD)
- Average absolute deviation (or simply called average deviation)
- Distance standard deviation

These are frequently used (together with scale factors) as estimators of scale parameters, in which capacity they are called estimates of scale. Robust measures of scale are those unaffected by a small number of outliers, and include the IQR and MAD.

All the above measures of statistical dispersion have the useful property that they are location-invariant and linear in scale. This means that if a random variable X has a dispersion of S_X then a linear transformation $Y = aX + b$ for real a and b should have dispersion $S_Y = |a|S_X$, where $|a|$is the absolute value of a, that is, ignores a preceding negative sign –.

Other measures of dispersion are dimensionless. In other words, they have no units even if the variable itself has units. These include:

- Coefficient of variation
- Quartile coefficient of dispersion
- Relative mean difference, equal to twice the Gini coefficient
- Entropy: While the entropy of a discrete variable is location-invariant and scale-independent, and therefore not a measure of dispersion in the above sense, the entropy of a continuous variable is location invariant and additive in scale: If Hz is the entropy of continuous variable z and y=ax+b, then Hy=Hx+log(a).

There are other measures of dispersion:

- Variance (the square of the standard deviation) – location-invariant but not linear in scale.
- Variance-to-mean ratio – mostly used for count data when the term coefficient

of dispersion is used and when this ratio is dimensionless, as count data are themselves dimensionless, not otherwise.

Some measures of dispersion have specialized purposes, among them the Allan variance and the Hadamard variance.

For categorical variables, it is less common to measure dispersion by a single number. One measure that does so is the discrete entropy.

Sources of Statistical Dispersion

In the physical sciences, such variability may result from random measurement errors: instrument measurements are often not perfectly precise, i.e., reproducible, and there is additional inter-rater variability in interpreting and reporting the measured results. One may assume that the quantity being measured is stable, and that the variation between measurements is due to observational error. A system of a large number of particles is characterized by the mean values of a relatively few number of macroscopic quantities such as temperature, energy, and density. The standard deviation is an important measure in Fluctuation theory, which explains many physical phenomena, including why the sky is blue.

In the biological sciences, the quantity being measured is seldom unchanging and stable, and the variation observed might additionally be intrinsic to the phenomenon: It may be due to inter-individual variability, that is, distinct members of a population differing from each other. Also, it may be due to intra-individual variability, that is, one and the same subject differing in tests taken at different times or in other differing conditions. Such types of variability are also seen in the arena of manufactured products; even there, the meticulous scientist finds variation.

In economics, finance, and other disciplines, regression analysis attempts to explain the dispersion of a dependent variable, generally measured by its variance, using one or more independent variables each of which itself has positive dispersion. The fraction of variance explained is called the coefficient of determination.

A Partial Ordering of Dispersion

A mean-preserving spread (MPS) is a change from one probability distribution A to another probability distribution B, where B is formed by spreading out one or more portions of A's probability density function while leaving the mean (the expected value) unchanged. The concept of a mean-preserving spread provides a partial ordering of probability distributions according to their dispersions: of two probability distributions, one may be ranked as having more dispersion than the other, or alternatively neither may be ranked as having more dispersion.

Shape of a Probability Distribution

In statistics, the concept of the shape of a probability distribution arises in questions of finding an appropriate distribution to use to model the statistical properties of a population, given a sample from that population. The shape of a distribution may be considered either descriptively, using terms such as "J-shaped", or numerically, using quantitative measures such as skewness and kurtosis.

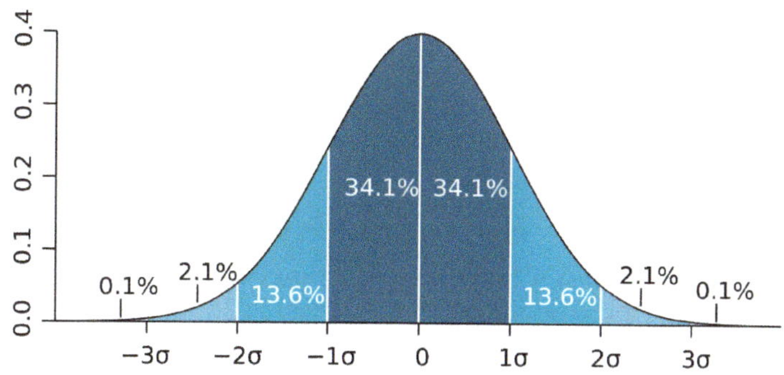

The normal distribution, often called the "bell curve"

Considerations of the shape of a distribution arise in statistical data analysis, where simple quantitative descriptive statistics and plotting techniques such as histograms can lead on to the selection of a particular family of distributions for modelling purposes.

Descriptions of Shape

The shape of a distribution will fall somewhere in a continuum where a flat distribution might be considered central and where types of departure from this include: mounded (or unimodal), U-shaped, J-shaped, reverse-J shaped and multi-modal. A bimodal distribution would have two high points rather than one. The shape of a distribution is sometimes characterized by the behaviors of the tails (as in a long or short tail). For example, a flat distribution can be said either to have no tails, or to have short tails. A normal distribution is usually regarded as having short tails, while an exponential distribution has exponential tails and a Pareto distribution has long tails. HHH

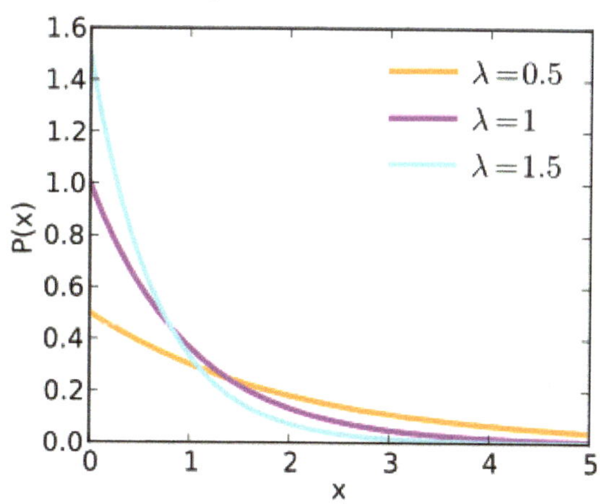

Exponential distribution

REFERENCES

1. Mann, Prem S. (1995). Introductory Statistics (2nd ed.). Wiley. ISBN 0-471-31009-3.
2. Dodge, Y. (2003). The Oxford Dictionary of Statistical Terms. OUP. ISBN. 0-19-850994-4.
3. Investopedia, Descriptive Statistics Terms
4. Trochim, William M. K. (2006). "Descriptive statistics". Research Methods Knowledge Base. Retrieved 14 March 2011.
5. Babbie, Earl R. (2009). The Practice of Social Research (12th ed.). Wadsworth. pp. 436–440. ISBN 0-495-59841-0.
6. Nick, Todd G. (2007). "Descriptive Statistics". Topics in Biostatistics. Methods in Molecular Biology. 404. New York: Springer. pp. 33–52. doi:10.1007/978-1-59745-530-5_3. ISBN 978-1-58829-531-6.
7. Weisberg H.F (1992) Central Tendency and Variability, Sage University Paper Series on Quantitative Applications in the Social Sciences, ISBN 0-8039-4007-6 p.2
8. Upton, G.; Cook, I. (2008) Oxford Dictionary of Statistics, OUP ISBN 978-0-19-954145-4 (entry for "central tendency")
9. Dodge, Y. (2003) The Oxford Dictionary of Statistical Terms, OUP for International Statistical Institute. ISBN 0-19-920613-9 (entry for "central tendency")
10. Johnson NL, Rogers CA (1951) "The moment problem for unimodal distributions". Annals of Mathematical Statistics, 22 (3) 433–439
11. Hotelling H, Solomons LM (1932) The limits of a measure of skewness. Annals Math Stat 3, 141–114
12. Garver (1932) Concerning the limits of a mesuare of skewness. Ann Math Stats 3(4) 141–142
13. Weisstein, Eric W. "Statistical Median". MathWorld.
14. Simon, Laura J.; "Descriptive statistics", Statistical Education Resource Kit, Pennsylvania State Department of Statistics
15. David J. Sheskin (27 August 2003). Handbook of Parametric and Nonparametric Statistical Procedures: Third Edition. CRC Press. pp. 7–. ISBN 978-1-4200-3626-8. Retrieved 25 February 2013.
16. Derek Bissell (1994). Statistical Methods for Spc and Tqm. CRC Press. pp. 26–. ISBN 978-0-412-39440-9. Retrieved 25 February 2013.
17. "Journal of Statistics Education, v13n2: Paul T. von Hippel". amstat.org.
18. Robson, Colin (1994). Experiment, Design and Statistics in Psychology. Penguin. pp. 42–45. ISBN 0-14-017648-9.
19. "AP Statistics Review - Density Curves and the Normal Distributions". Retrieved 16 March 2015.
20. Newman, Mark EJ. "Power laws, Pareto distributions and Zipf's law."

Contemporary physics 46.5 (2005): 323–351.

21. Stroock, Daniel (2011). Probability Theory. Cambridge University Press. p. 43. ISBN 978-0-521-13250-3.
22. "An Error Occurred Setting Your User Cookie". siam.org.
23. Mallows, Colin (August 1991). "Another comment on O'Cinneide". The American Statistician. 45 (3): 257. doi:10.1080/00031305.1991.10475815.
24. Piché, Robert (2012). Random Vectors and Random Sequences. Lambert Academic Publishing. ISBN 978-3659211966.
25. Kemperman, Johannes H. B. (1987). Dodge, Yadolah, ed. "The median of a finite measure on a Banach space: Statistical data analysis based on the L1-norm and related methods". Papers from the First International Conference held at Neuchâtel, August 31–September 4, 1987. Amsterdam: North-Holland Publishing Co.: 217–230. MR 949228.
26. Milasevic, Philip; Ducharme, Gilles R. (1987). "Uniqueness of the spatial median". Annals of Statistics. 15 (3): 1332–1333. doi:10.1214/aos/1176350511. MR 902264.
27. Merkle, M. (2005). "Jensen's inequality for medians". Statistics & Probability Letters. 71 (3): 277–281. doi:10.1016/j.spl.2004.11.010.
28. Bentley, Jon L.; McIlroy, M. Douglas (1993). "Engineering a sort function". Software—Practice and Experience. 23 (11): 1249–1265. doi:10.1002/spe.4380231105.
29. Rousseeuw, Peter J.; Bassett, Gilbert W., Jr. (1990). "The remedian: a robust averaging method for large data sets" (PDF). J. Amer. Stat. Soc. 85 (409): 97–104.
30. Stigler, Stephen (December 1973). "Studies in the History of Probability and Statistics. XXXII: Laplace, Fisher and the Discovery of the Concept of Sufficiency". Biometrika. 60 (3): 439–445. doi:10.1093/biomet/60.3.439. JSTOR 2334992. MR 0326872.
31. Rider, Paul R. (1960). "Variance of the median of small samples from several special populations". J. Amer. Statist. Assoc. 55 (289): 148–150. doi:10.1080/01621459.1960.10482056.
32. Stuart, Alan; Ord, Keith (1994). Kendall's Advanced Theory of Statistics. London: Arnold. ISBN 0340614307.
33. Efron, B. (1982). The Jackknife, the Bootstrap and other Resampling Plans. Philadelphia: SIAM. ISBN 0898711797.
34. Shao, J.; Wu, C. F. (1989). "A General Theory for Jackknife Variance Estimation". Ann. Stat. 17 (3): 1176–1197. doi:10.1214/aos/1176347263. JSTOR 2241717.
35. Efron, B. (1979). "Bootstrap Methods: Another Look at the Jackknife". Ann. Stat. 7 (1): 1–26. doi:10.1214/aos/1176344552. JSTOR 2958830.
36. Hall, P.; Martin, M. A. (1988). "Exact Convergence Rate of Bootstrap Quantile Variance Estimator". Probab Theory Related Fields. 80 (2): 261–268. doi:10.1007/BF00356105.

37. Jiménez-Gamero, M. D.; Munoz-García, J.; Pino-Mejías, R. (2004). "Reduced bootstrap for the median". Statistica Sinica. 14 (4): 1179–1198.
38. Kenney, J. F.; Keeping, E. S. (1962). "The Median". Mathematics of Statistics, Pt. 1 (3rd ed.). Princeton, NJ: Van Nostrand. pp. 211–212.
39. Hettmansperger, Thomas P.; McKean, Joseph W. (1998). Robust nonparametric statistical methods. Kendall's Library of Statistics. 5. London: Edward Arnold. ISBN 0-340-54937-8. MR 1604954.
40. Bonett DG, Seier E (2006) Confidence interval for a coefficient of dispersion in non-normal distributions. Biometrical Journal 48 (1) 144–148
41. Statistical Calculation Definitions for Mass Appraisal (PDF). Iowa.gov. Archived from the original (PDF) on 11 November 2010. Median Ratio: The ratio located midway between the highest ratio and the lowest ratio when individual ratios for a class of realty are ranked in ascending or descending order. The median ratio is most frequently used to determine the level of assessment for a given class of real estate.
42. Assessment equity in New York: Results from the 2010 market value survey. Archived from the original on 6 November 2012.
43. Summary of the Assessment Process (PDF). state.sd.us. South Dakota Department of Revenue - Property/Special Taxes Division. Archived from the original (PDF) on 10 May 2009.
44. Small, Christopher G. A survey of multidimensional medians. International Statistical Review/Revue Internationale de Statistique (1990): 263–277. DOI: 10.2307/1403809 Stable URL: http://www.jstor.org/stable/1403809
45. Niinimaa, A., and H. Oja. Multivariate median. Encyclopedia of statistical sciences (1999).
46. Mosler, Karl. Multivariate Dispersion, Central Regions, and Depth: The Lift Zonoid Approach. Vol. 165. Springer Science & Business Media, 2012.
47. Puri, Madan L.; Sen, Pranab K.; Nonparametric Methods in Multivariate Analysis, John Wiley & Sons, New York, NY, 197l. (Reprinted by Krieger Publishing)
48. Wesolowsky, G. (1993). The Weber problem: History and perspective. Location Science. 1: 5–23.
49. Oja, Hannu (2010). Multivariate nonparametric methods with R: An approach based on spatial signs and ranks. Lecture Notes in Statistics. 199. New York, NY: Springer. pp. xiv+232. doi:10.1007/978-1-4419-0468-3. ISBN 978-1-4419-0467-6. MR 2598854.
50. Vardi, Yehuda; Zhang, Cun-Hui (2000). The multivariate l1-median and associated data depth. Proceedings of the National Academy of Sciences of the United States of America. 97 (4): 1423–1426. doi:10.1073/pnas.97.4.1423.
51. Lopuhaä, Hendrick P.; Rousseeuw, Peter J. (1991). Breakdown points of affine equivariant estimators of multivariate location and covariance matrices. Annals of Statistics. 19 (1): 229–248. doi:10.1214/aos/1176347978. JSTOR 2241852.
52. Wilcox, Rand R. (2001), Theil–Sen estimator, Fundamentals of Modern Statistical

Methods: Substantially Improving Power and Accuracy, Springer-Verlag, pp. 207–210, ISBN 978-0-387-95157-7.

53. Wald, A. (1940). The Fitting of Straight Lines if Both Variables are Subject to Error. Annals of Mathematical Statistics. 11 (3): 282–300. doi:10.1214/aoms/1177731868. JSTOR 2235677.
54. Nair, K. R.; Shrivastava, M. P. (1942). On a Simple Method of Curve Fitting. Sankhyā: The Indian Journal of Statistics. 6 (2): 121–132. JSTOR 25047749.
55. Brown, G. W.; Mood, A. M. (1951). On Median Tests for Linear Hypotheses. Proc Second Berkeley Symposium on Mathematical Statistics and Probability. Berkeley, CA: University of California Press. pp. 159–166. Zbl 0045.08606.
56. Tukey, J. W. (1977). Exploratory Data Analysis. Reading, MA: Addison-Wesley. ISBN 0201076160.
57. Brown, George W. (1947). On Small-Sample Estimation. Annals of Mathematical Statistics. 18 (4): 582–585. doi:10.1214/aoms/1177730349. JSTOR 2236236.
58. Lehmann, Erich L. (1951). A General Concept of Unbiasedness. Annals of Mathematical Statistics. 22 (4): 587–592. doi:10.1214/aoms/1177729549. JSTOR 2236928.
59. Birnbaum, Allan (1961). A Unified Theory of Estimation, I. Annals of Mathematical Statistics. 32 (1): 112–135. doi:10.1214/aoms/1177705145. JSTOR 2237612.
60. van der Vaart, H. Robert (1961). Some Extensions of the Idea of Bias. Annals of Mathematical Statistics. 32 (2): 436–447. doi:10.1214/aoms/1177705051. JSTOR 2237754. MR 125674.
61. Pfanzagl, Johann; with the assistance of R. Hamböker (1994). Parametric Statistical Theory. Walter de Gruyter. ISBN 3-11-013863-8. MR 1291393.
62. Pfanzagl, Johann. On optimal median unbiased estimators in the presence of nuisance parameters. The Annals of Statistics (1979): 187–193.
63. Brown, L. D.; Cohen, Arthur; Strawderman, W. E. A Complete Class Theorem for Strict Monotone Likelihood Ratio With Applications. Ann. Statist. 4 (1976), no. 4, 712–722. doi:10.1214/aos/1176343543. http://projecteuclid.org/euclid.aos/1176343543.
64. Page 713: Brown, L. D.; Cohen, Arthur; Strawderman, W. E. A Complete Class Theorem for Strict Monotone Likelihood Ratio With Applications. Ann. Statist. 4 (1976), no. 4, 712–722. doi:10.1214/aos/1176343543. http://projecteuclid.org/euclid.aos/1176343543.
65. Talmud and Modern Economics
66. Modern Economic Theory in the Talmud by Yisrael Aumann
67. Stigler, S. M. (1986). The History of Statistics: The Measurement of Uncertainty Before 1900. Harvard University Press. ISBN 0674403401.
68. Jaynes, E.T. (2007). Probability theory : the logic of science (5. print. ed.). Cambridge [u.a.]: Cambridge Univ. Press. p. 172. ISBN 978-0-521-59271-0.

69. Laplace PS de (1818) Deuxième supplément à la Théorie Analytique des Probabilités, Paris, Courcier

70. Keynes, J.M. (1921) A Treatise on Probability. Pt II Ch XVII §5 (p 201) (2006 reprint, Cosimo Classics, ISBN 9781596055308 : multiple other reprints)

71. Galton F (1881) Report of the Anthropometric Committee pp 245–260. Report of the 51st Meeting of the British Association for the Advancement of Science

CHAPTER 3

Data Collection

Chapter Contents

- **Introduction**
- **Types**
- **Study Design**
- **Survey Methodology**

Data collection is the systematic approach to gathering and measuring information from a variety of sources to get a complete and accurate picture of an area of interest. Data collection enables a person or organization to answer relevant questions, evaluate outcomes and make predictions about future probabilities and trends.

INTRODUCTION

Data collection is the process of gathering and measuring information on targeted variables in an established systematic fashion, which then enables one to answer relevant questions and evaluate outcomes. The data collection component of research is common to all fields of study including physical and social sciences, humanities and business. It help us to collect the main points as gathered information. While methods vary by discipline, the emphasis on ensuring accurate and honest collection remains the same. The goal for all data collection is to capture quality evidence that then translates to rich data analysis and allows the building of a convincing and credible answer to questions that have been posed.

Adélie penguins are identified and weighed each time they cross the automated weighbridge on their way to or from the sea.

Importance

Regardless of the field of study or preference for defining data (quantitative or qualitative), accurate data collection is essential to maintaining the integrity of research. Both the selection of appropriate data collection instruments (existing, modified, or newly developed) and clearly delineated instructions for their correct use reduce the likelihood of errors occurring.

A formal data collection process is necessary as it ensures that the data gathered are both defined and accurate and that subsequent decisions based on arguments embodied in the findings are valid. The process provides both a baseline from which to measure and in certain cases a target on what to improve.

TYPES

Generally there are three types of data collection and they are:

1. Surveys: Standardized paper-and-pencil or phone questionnaires that ask predetermined questions.
2. Interviews: Structured or unstructured one-on-one directed conversations with key individuals or leaders in a community.
3. Focus groups: Structured interviews with small groups of like individuals using standardized questions, follow-up questions, and exploration of other topics that arise to better understand participants.

Consequences from improperly collected data include:

- Inability to answer research questions accurately;
- Inability to repeat and validate the study.

Impact of Faulty Data

Distorted findings result in wasted resources and can mislead other researchers to pursue fruitless avenues of investigation. This compromises decisions for public policy.

While the degree of impact from faulty data collection may vary by discipline and the nature of investigation, there is the potential to cause disproportionate harm when these research results are used to support public policy recommendations.

STUDY DESIGN

Statistical Population

In statistics, a population is a set of similar items or events which is of interest for some question or experiment. A statistical population can be a group of actually existing objects (e.g. the set of all stars within the Milky Way galaxy) or a hypothetical and potentially infinite group of objects conceived as a generalization from experience (e.g. the set of all possible hands in a game of poker). A common aim of statistical analysis is to produce information about some chosen population.

In statistical inference, a subset of the population (a statistical sample) is chosen to represent the population in a statistical analysis. If a sample is chosen properly, characteristics of the entire population that the sample is drawn from can be estimated from corresponding characteristics of the sample.

Subpopulation

A subset of a population that shares one or more additional properties is called a subpopulation. For example, if the population is all Egyptian people, a subpopulation is all Egyptian males; if the population is all pharmacies in the world, a subpopulation is all pharmacies in Egypt. By contrast, a sample is a subset of a population that is not chosen to share any additional property.

Descriptive statistics may yield different results for different subpopulations. For instance, a particular medicine may have different effects on different subpopulations, and these effects may be obscured or dismissed if such special subpopulations are not identified and examined in isolation.

Similarly, one can often estimate parameters more accurately if one separates out subpopulations: the distribution of heights among people is better modeled by considering men and women as separate subpopulations, for instance.

Populations consisting of subpopulations can be modeled by mixture models, which combine the distributions within subpopulations into an overall population distribution. Even if subpopulations are well-modeled by given simple models, the overall population may be poorly fit by a given simple model – poor fit may be evidence for existence of subpopulations. For example, given two equal subpopulations, both normally distributed, if they have the same standard deviation and different means, the overall distribution will exhibit low kurtosis relative to a single normal distribution – the means of the subpopulations fall on the shoulders of the overall distribution. If sufficiently separated, these form a bimodal distribution, otherwise it simply has a wide peak. Further, it will exhibit overdispersion relative to a single normal distribution with the given variation. Alternatively, given two subpopulations with the same mean and different standard deviations, the overall population will exhibit high kurtosis, with a sharper peak and heavier tails (and correspondingly shallower shoulders) than a single distribution.

Statistic

A statistic (singular) or sample statistic is a single measure of some attribute of a sample (e.g., its arithmetic mean value). It is calculated by applying a function (statistical algorithm) to the values of the items of the sample, which are known together as a set of data.

More formally, statistical theory defines a statistic as a function of a sample where the function itself is independent of the sample's distribution; that is, the function can be stated before realization of the data. The term statistic is used both for the function and for the value of the function on a given sample.

A statistic is distinct from a statistical parameter, which is not computable because often the population is much too large to examine and measure all its items. However, a statistic, when used to estimate a population parameter, is called an estimator. For instance, the sample mean is a statistic that estimates the population mean, which is a parameter.

When a statistic (a function) is being used for a specific purpose, it may be referred to by a name indicating its purpose: in descriptive statistics, a descriptive statistic is used to describe the data; in estimation theory, an estimator is used to estimate a parameter of the distribution (population); in statistical hypothesis testing, a test statistic is used to test a hypothesis. However, a single statistic can be used for multiple purposes – for example the sample mean can be used to describe a data set, to estimate the population mean, or to test a hypothesis.

Examples

In calculating the arithmetic mean of a sample, for example, the algorithm works by summing all the data values observed in the sample and then dividing this sum by the number of data items. This single measure, the mean of the sample, is called a statistic; its value is frequently used as an estimate of the mean value of all items comprising the population from which the sample is drawn. The population mean is also a single measure; however, it is not called a statistic, because it is not obtained from a sample; instead it is called a population parameter, because it is obtained from the whole population.

Other examples of statistics include

- Sample mean discussed in the example above and sample median
- Sample variance and sample standard deviation
- Sample quantiles besides the median, e.g., quartiles and percentiles
- Test statistics, such as t statistics, chi-squared statistics, f statistics
- Order statistics, including sample maximum and minimum

- Sample moments and functions thereof, including kurtosis and skewness
- Various functionals of the empirical distribution function

Properties

Observability

A statistic is an observable random variable, which differentiates it both from a parameter that is a generally unobservable quantity describing a property of a statistical population, and from an unobservable random variable, such as the difference between an observed measurement and a population average. A parameter can only be computed exactly if the entire population can be observed without error; for instance, in a perfect census or for a population of standardized test takers.

Statisticians often contemplate a parameterized family of probability distributions, any member of which could be the distribution of some measurable aspect of each member of a population, from which a sample is drawn randomly. For example, the parameter may be the average height of 25-year-old men in North America. The height of the members of a sample of 100 such men are measured; the average of those 100 numbers is a statistic. The average of the heights of all members of the population is not a statistic unless that has somehow also been ascertained (such as by measuring every member of the population). The average height that would be calculated using the all of the individual heights of all 25-year-old North American men is a parameter, and not a statistic.

Statistical Properties

Important potential properties of statistics include completeness, consistency, sufficiency, unbiasedness, minimum mean square error, low variance, robustness, and computational convenience.

Information of a Statistic

Information of a statistic on model parameters can be defined in several ways. The most common is the Fisher information, which is defined on the statistic model induced by the statistic. Kullback information measure can also be used.

Effect Size

In statistics, an effect size is a quantitative measure of the strength of a phenomenon. Examples of effect sizes are the correlation between two variables, the regression coefficient in a regression, the mean difference, or even the risk with which something happens, such as how many people survive after a heart attack for every one person that does not survive. For each type of effect size, a larger absolute value always indicates a stronger effect. Effect sizes complement statistical hypothesis testing, and play an important role in power analyses, sample size planning, and in meta-analyses. They are the first item (magnitude) in the MAGIC criteria for evaluating the strength of a statistical claim.

Especially in meta-analysis, where the purpose is to combine multiple effect sizes, the standard error (S.E.) of the effect size is of critical importance. The S.E. of the effect size is used to weight effect sizes when combining studies, so that large studies are considered more important than small studies in the analysis. The S.E. of the effect size is calculated differently for each type of effect size, but generally only requires knowing the study's sample size (N), or the number of observations in each group (n's).

Reporting effect sizes is considered good practice when presenting empirical research findings in many fields. The reporting of effect sizes facilitates the interpretation of the substantive, as opposed to the statistical, significance of a research result. Effect sizes are particularly prominent in social and medical research. Relative and absolute measures of effect size convey different information, and can be used complementarily. A prominent task force in the psychology research community made the following recommendation:

> Always present effect sizes for primary outcomes...If the units of measurement are meaningful on a practical level (e.g., number of cigarettes smoked per day), then we usually prefer an unstandardized measure (regression coefficient or mean difference) to a standardized measure (r or d).
>
> —L. Wilkinson and APA Task Force on Statistical Inference (1999, p. 599)

Overview

Population and Sample Effect Sizes

The term effect size can refer to the value of a statistic calculated from a sample of data, the value of a parameter of a hypothetical statistical population, or to the equation that operationalizes how statistics or parameters lead to the effect size value. Conventions for distinguishing sample from population effect sizes follow standard statistical practices — one common approach is to use Greek letters like ρ to denote population parameters and Latin letters like r to denote the corresponding statistic; alternatively, a "hat" can be placed over the population parameter to denote the statistic, e.g. with $\hat{\rho}$ being the estimate of the parameter ρ.

As in any statistical setting, effect sizes are estimated with sampling error, and may be biased unless the effect size estimator that is used is appropriate for the manner in which the data were sampled and the manner in which the measurements were made. An example of this is publication bias, which occurs when scientists only report results when the estimated effect sizes are large or are statistically significant. As a result, if many researchers carry out studies with low statistical power, the reported effect sizes will tend to be larger than the true (population) effects, if any. Another example where effect sizes may be distorted is in a multiple trial experiment, where the effect size calculation is based on the averaged or aggregated response across the trials.

Relationship to Test Statistics

Sample-based effect sizes are distinguished from test statistics used in hypothesis testing, in that they estimate the strength (magnitude) of, for example, an apparent relationship, rather than assigning a significance level reflecting whether the magnitude of the relationship observed could be due to chance. The effect size does not directly determine the significance level, or vice versa. Given a sufficiently large sample size, a non-null statistical comparison will always show a statistically significant results unless the population effect size is exactly zero (and even there it will show statistical significance at the rate of the Type I error used). For example, a sample Pearson correlation coefficient of 0.01 is statistically significant if the sample size is 1000. Reporting only the significant p-value from this analysis could be misleading if a correlation of 0.01 is too small to be of interest in a particular application.

Standardized and Unstandardized Effect Sizes

The term effect size can refer to a standardized measure of effect (such as r, Cohen's d, or the odds ratio), or to an unstandardized measure (e.g., the difference between group means or the unstandardized regression coefficients). Standardized effect size measures are typically used when:

- the metrics of variables being studied do not have intrinsic meaning (e.g., a score on a personality test on an arbitrary scale),
- results from multiple studies are being combined,
- some or all of the studies use different scales, or
- it is desired to convey the size of an effect relative to the variability in the population.

In meta-analyses, standardized effect sizes are used as a common measure that can be calculated for different studies and then combined into an overall summary.

Types

About 50 to 100 different measures of effect size are known.

Correlation family: Effect sizes based on "variance explained"

These effect sizes estimate the amount of the variance within an experiment that is "explained" or "accounted for" by the experiment's model.

Pearson r or Correlation Coefficient

Pearson's correlation, often denoted r and introduced by Karl Pearson, is widely used as an effect size when paired quantitative data are available; for instance if one were studying the relationship between birth weight and longevity. The correlation coefficient can also be used when the data are binary. Pearson's r can vary in magnitude from −1 to 1, with −1 indicating a perfect negative linear relation, 1 indicating a perfect positive linear relation, and 0 indicating no linear relation between two variables. Cohen gives the following guidelines for the social sciences:

Effect size	*r*
Small	0.10

Medium	0.30
Large	0.50

Coefficient of Determination

A related effect size is r^2, the coefficient of determination (also referred to as R^2 or "r-squared"), calculated as the square of the Pearson correlation r. In the case of paired data, this is a measure of the proportion of variance shared by the two variables, and varies from 0 to 1. For example, with an r of 0.21 the coefficient of determination is 0.0441, meaning that 4.4% of the variance of either variable is shared with the other variable. The r^2 is always positive, so does not convey the direction of the correlation between the two variables.

Eta-squared (η^2)

Eta-squared describes the ratio of variance explained in the dependent variable by a predictor while controlling for other predictors, making it analogous to the r^2. Eta-squared is a biased estimator of the variance explained by the model in the population (it estimates only the effect size in the sample). This estimate shares the weakness with r^2 that each additional variable will automatically increase the value of η^2. In addition, it measures the variance explained of the sample, not the population, meaning that it will always overestimate the effect size, although the bias grows smaller as the sample grows larger.

$$\eta^2 = \frac{SS_{\text{Treatment}}}{SS_{\text{Total}}}.$$

Omega-squared (ω^2)

A less biased estimator of the variance explained in the population is ω^2

$$\omega^2 = \frac{SS_{\text{treatment}} - df_{\text{treatment}} * MS_{\text{error}}}{SS_{\text{total}} + MS_{\text{error}}}.$$

This form of the formula is limited to between-subjects analysis with equal sample sizes in all cells. Since it is less biased (although not unbiased), ω^2 is preferable to η^2; however, it can be more inconvenient to calculate for complex analyses. A generalized form of the estimator has been published for between-subjects and within-subjects analysis, repeated measure, mixed design, and randomized block design experiments. In addition, methods to calculate partial Omega^2 for individual factors and combined factors in designs with up to three independent variables have been published.

Cohen's f^2

Cohen's f^2 is one of several effect size measures to use in the context of an F-test for ANOVA or multiple regression. Its amount of bias (overestimation of the effect size for the ANOVA) depends on the bias of its underlying measurement of variance explained (e.g., R^2, η^2, ω^2).

The f^2 effect size measure for multiple regression is defined as:

$$f^2 = \frac{R^2}{1 - R^2}$$

where R^2 is the squared multiple correlation.

Likewise, f^2 can be defined as:

$$f^2 = \frac{\eta^2}{1 - \eta^2} \text{ or } f^2 = \frac{\omega^2}{1 - \omega^2}$$

for models described by those effect size measures.

The f^2 effect size measure for sequential multiple regression and also common for PLS modeling is defined as:

$$f^2 = \frac{R^2_{AB} - R^2_A}{1 - R^2_{AB}}$$

where R^2_A is the variance accounted for by a set of one or more independent variables A, and R^2_{AB} is the combined variance accounted for by A and another set of one or more independent variables of interest B. By convention, f^2_B effect sizes of 0.02, 0.15, and 0.35 are termed small, medium, and large, respectively.

Cohen's $\hat{f}$ can also be found for factorial analysis of variance (ANOVA) working backwards using :

$$\hat{f}_{\text{effect}} = \sqrt{(df_{\text{effect}}/N)(F_{\text{effect}} - 1)}.$$

In a balanced design (equivalent sample sizes across groups) of ANOVA, the corresponding population parameter of f^2 is

$$\frac{SS(\mu_1, \mu_2, \ldots, \mu_K)}{K \times \sigma^2},$$

wherein μ_j denotes the population mean within the jth group of the total K groups, and σ the equivalent population standard deviations within each groups. SS is the sum of squares in ANOVA.

Cohen's q

Another measure that is used with correlation differences is Cohen's q. This is the difference between two Fisher transformed Pearson regression coefficients. In symbols this is

$$q = \frac{1}{2}\log\frac{1+r_1}{1-r_1} - \frac{1}{2}\log\frac{1+r_2}{1-r_2}$$

where r_1 and r_2 are the regressions being compared. The expected value of q is zero and its variance is

$$var(q) = \frac{1}{N_1 - 3} + \frac{1}{N_2 - 3}$$

where N_1 and N_2 are the number of data points in the first and second regression respectively.

Difference family: Effect sizes based on differences between means

A (population) effect size θ based on means usually considers the standardized mean difference between two populations

$$\theta = \frac{\mu_1 - \mu_2}{\sigma},$$

where μ_1 is the mean for one population, μ_2 is the mean for the other population, and σ is a standard deviation based on either or both populations.

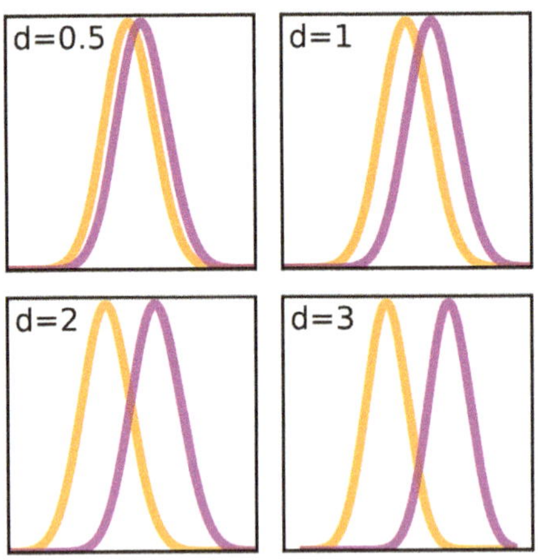

Plots of Gaussian densities illustrating various values of Cohen's d.

In the practical setting the population values are typically not known and must be estimated from sample statistics. The several versions of effect sizes based on means differ with respect to which statistics are used.

This form for the effect size resembles the computation for a t-test statistic, with the critical difference that the t-test statistic includes a factor of $\sqrt{n}$. This means that for a given effect size, the significance level increases with the sample size. Unlike the t-test statistic, the effect size aims to estimate a population parameter and is not affected by the sample size.

Cohen's d

Cohen's d is defined as the difference between two means divided by a standard deviation for the data, i.e.

$$d = \frac{\bar{x}_1 - \bar{x}_2}{s}.$$

Jacob Cohen defined s, the pooled standard deviation, as (for two independent samples):[:67]

$$s = \sqrt{\frac{(n_1 - 1)s_1^2 + (n_2 - 1)s_2^2}{n_1 + n_2 - 2}}$$

where the variance for one of the groups is defined as

$$s_1^2 = \frac{1}{n_1 - 1}\sum_{i=1}^{n_1}(x_{1,i} - \bar{x}_1)^2,$$

and similar for the other group.

The table below contains descriptors for magnitudes of d = 0.01 to 2.0, as initially suggested by Cohen and expanded by Sawilowsky.

Effect Size	d	**Reference**
Very Small	0.01	Sawilowsky, 2009
Small	0.20	Cohen, 1988
Medium	0.50	Cohen, 1988
Large	0.80	Cohen, 1988
Very Large	1.20	Sawilowsky, 2009
Huge	2.0	Sawilowsky, 2009

Other authors choose a slightly different computation of the standard deviation when referring to "Cohen's d" where the denominator is without "-2"

$$s = \sqrt{\frac{(n_1 - 1)s_1^2 + (n_2 - 1)s_2^2}{n_1 + n_2}}$$

This definition of "Cohen's d" is termed the maximum likelihood estimator by Hedges and Olkin, and it is related to Hedges' g by a scaling factor (see below).

So, in the example above of visiting England and observing men's and women's heights, the data (Aaron,Kromrey,& Ferron, 1998, November; from a 2004 UK representative sample of 2436 men and 3311 women) are:

- Men: mean height = 1750 mm; standard deviation = 89.93 mm
- Women: mean height = 1612 mm; standard deviation = 69.05 mm

The effect size (using Cohen's d) would equal 1.72 (95% confidence intervals: 1.66 – 1.78). This is very large and you should have no problem in detecting that there is a consistent height difference, on average, between men and women.

With two paired samples, we look at the distribution of the difference scores. In that case, s is the standard deviation of this distribution of difference scores. This creates the following relationship between the t-statistic to test for a difference in the means of the two groups and Cohen's d:

$$t = \frac{\bar{X}_1 - \bar{X}_2}{SE} = \frac{\bar{X}_1 - \bar{X}_2}{\frac{SD}{\sqrt{N}}} = \frac{\sqrt{N}(\bar{X}_1 - \bar{X}_2)}{SD}$$

and

$$d = \frac{\bar{X}_1 - \bar{X}_2}{SD} = \frac{t}{\sqrt{N}}$$

Cohen's d is frequently used in estimating sample sizes for statistical testing. A lower Cohen's d indicates the necessity of larger sample sizes, and vice versa, as can subsequently be determined together with the additional parameters of desired significance level and statistical power.

Glass' **Δ**

In 1976 Gene V. Glass proposed an estimator of the effect size that uses only the standard deviation of the second group

$$\Delta = \frac{\bar{x}_1 - \bar{x}_2}{s_2}$$

The second group may be regarded as a control group, and Glass argued that if several treatments were compared to the control group it would be better to use just the standard deviation computed from the control group, so that effect sizes would not differ under equal means and different variances.

Under a correct assumption of equal population variances a pooled estimate for σ is more precise.

Hedges' g

Hedges' g, suggested by Larry Hedges in 1981, is like the other measures based on a standardized difference[79]

$$g = \frac{\bar{x}_1 - \bar{x}_2}{s^*}$$

where the pooled standard deviation s* is computed as:

$$s^* = \sqrt{\frac{(n_1-1)s_1^2 + (n_2-1)s_2^2}{n_1+n_2-2}}.$$

However, as an estimator for the population effect size θ it is biased. Nevertheless, this bias can be approximately corrected through multiplication by a factor

$$g^* = J(n_1+n_2-2)\; g \approx \left(1 - \frac{3}{4(n_1+n_2)-9}\right)\; g$$

Hedges and Olkin refer to this less-biased estimator g* as d, but it is not the same as Cohen's d. The exact form for the correction factor J() involves the gamma function[104]

$$J(a) = \frac{\Gamma(a/2)}{\sqrt{a/2}\,\Gamma((a-1)/2)}.$$

Ψ, Root-Mean-Square Standardized Effect

A similar effect size estimator for multiple comparisons (e.g., ANOVA) is the Ψ root-mean-square standardized effect. This essentially presents the omnibus difference of the entire model adjusted by the root mean square, analogous to d or g. The simplest formula for Ψ, suitable for one-way ANOVA, is

$$\Psi = \sqrt{\left(\frac{1}{k-1}\right)\frac{\Sigma(\bar{x}_j - \bar{X})^2}{MS_{error}}}$$

In addition, a generalization for multi-factorial designs has been provided.

Distribution of Effect Sizes Based on Means

Provided that the data is Gaussian distributed a scaled Hedges' g, $\sqrt{n_1 n_2/(n_1+n_2)}\,g$, follows a noncentral t-distribution with the noncentrality parameter $\sqrt{n_1 n_2/(n_1+n_2)}\theta$ and $(n_1 + n_2 - 2)$ degrees of freedom. Likewise, the scaled Glass' Δ is distributed with $n_2 - 1$ degrees of freedom.

From the distribution it is possible to compute the expectation and variance of the effect sizes.

In some cases large sample approximations for the variance are used. One suggestion for the variance of Hedges' unbiased estimator is[86]

$$\hat{\sigma}^2(g^*) = \frac{n_1+n_2}{n_1 n_2} + \frac{(g^*)^2}{2(n_1+n_2)}.$$

Categorical family: Effect sizes for Associations Among Categorical Variables

Commonly used measures of association for the chi-squared test are the Phi coefficient and Cramér's V (sometimes referred to as Cramér's phi and denoted as φ_c). Phi is related to the point-biserial correlation coefficient and Cohen's d and estimates the extent of the relationship between two variables (2 x 2). Cramér's V may be used with variables having more than two levels.

Phi can be computed by finding the square root of the chi-squared statistic divided by the sample size.

Similarly, Cramér's V is computed by taking the square root of the chi-squared statistic divided by the sample size and the length of the minimum dimension (k is the smaller of the number of rows r or columns c).

φ_c is the intercorrelation of the two discrete variables and may be computed for any value of r or c. However, as chi-squared values tend to increase with the number of cells, the greater the difference between r and c, the more likely V will tend to 1 without strong evidence of a meaningful correlation.

$\phi = \sqrt{\frac{\chi^2}{N}}$	$\phi_c = \sqrt{\frac{\chi^2}{N(k-1)}}$
Phi (φ)	**Cramér's V (φ_c)**

Cramér's V may also be applied to 'goodness of fit' chi-squared models (i.e. those where c=1). In this case it functions as a measure of tendency towards a single outcome (i.e. out of k outcomes). In such a case one must use r for k, in order to preserve the 0 to 1 range of V. Otherwise, using c would reduce the equation to that for Phi.

Cohen's w

Another measure of effect size used for chi square tests is Cohen's w. This is defined as

$$w = \sqrt{\sum_{i=1}^{N} \frac{(p_{0i} - p_{1i})^2}{p_{0i}}}$$

where p_{0i} is the value of the ith cell under H_0 and p_{1i} is the value of the ith cell under H_1.

Odds Ratio

The odds ratio (OR) is another useful effect size. It is appropriate when the research question focuses on the degree of association between two binary variables. For example, consider a study of spelling ability. In a control group, two students pass the class for everyone who fails, so the odds of passing are two to one (or 2/1 = 2). In the treatment group, six students pass for everyone who fails, so the odds of passing are six to one (or 6/1 = 6). The effect size can be computed by noting that the odds of passing in the treatment group are three times higher than in the control group (because 6 divided by 2 is 3). Therefore, the odds ratio is 3. Odds ratio statistics are on a different scale than Cohen's d, so this '3' is not comparable to a Cohen's d of 3.

Relative Risk

The relative risk (RR), also called risk ratio, is simply the risk (probability) of an event relative to some independent variable. This measure of effect size differs from the odds ratio in that it compares probabilities instead of odds, but asymptotically approaches the latter for small probabilities. Using the example above, the probabilities for those in the control group and treatment group passing is 2/3 (or 0.67) and 6/7 (or 0.86), respectively. The effect size can be computed the same as above, but using the probabilities instead. Therefore, the relative risk is 1.28. Since rather large probabilities of passing were used, there is a large difference between relative risk and odds ratio. Had failure (a smaller probability) been used as the event (rather than passing), the difference between the two measures of effect size would not be so great.

While both measures are useful, they have different statistical uses. In medical research, the odds ratio is commonly used for case-control studies, as odds, but not probabilities, are usually estimated. Relative risk is commonly used in randomized controlled trials and cohort studies, but relative risk contributes to overestimations of the effectiveness of interventions.

Risk Difference

The risk difference (RD), sometimes called absolute risk reduction, is simply the difference in risk (probability) of an event between two groups. It is a useful measure in experimental research, since RD tells you the extent to which an experimental interventions changes the probability of an event or outcome. Using the example above, the probabilities for those in the control group and treatment group passing is 2/3 (or 0.67) and 6/7 (or 0.86), respectively, and so the RD effect size is 0.86 - 0.67 = 0.19 (or 19%). RD is the superior measure for assessing effectiveness of interventions.

Cohen's h

One measure used in power analysis when comparing two independent proportions is Cohen's h. This is defined as follows

$$h = 2\left(\arcsin\sqrt{p_1} - \arcsin\sqrt{p_2}\right)$$

where p_1 and p_2 are the proportions of the two samples being compared and arcsin is the arcsine transformation.

Common Language Effect Size

To more easily describe the meaning of an effect size, to people outside statistics, the common language effect size, as the name implies, was designed to communicate it in plain English. It is used to describe a difference between two groups and was proposed, as well as named, by Kenneth McGraw and S. P. Wong in 1992. They used the following example (about heights of men and women): "in any random pairing of young adult males and females, the probability of the male being taller than the female is .92, or in simpler terms yet, in 92 out of 100 blind dates among young adults, the male will be taller than the female", when describing the population value of the common language effect size.

The population value, for the common language effect size, is often reported like this, in terms of pairs randomly chosen from the population. Kerby (2014) notes that a pair, defined as a score in one group paired with a score in another group, is a core concept of the common language effect size.

As an other example, consider a scientific study (maybe of a treatment for some chronic disease, such as arthritis) with ten people in the treatment group and ten people in a control group. If everyone in the treatment group is compared to everyone in the control group, then there are (10×10=) 100 pairs. At the end of the study, the outcome is rated into a score, for each individual (for example on a scale of mobility and pain, in the case of an arthritis study), and then all the scores are compared between the pairs. The results, as the percent of pairs that support the hypothesis, is the common language effect size. In the example study it could be (lets say) .80, if 80 out of the 100 comparison pairs shows a better outcome for the treatment group, than the control group, and the report may read as follows: "When a patient in the treatment group was compared to a patient in the control group, in 80 of 100 pairs the treated patient showed a better treatment outcome." The sample value, in for example a study like this, is an unbiased estimator of the population value.

Vargha and Delaney generalized the common language effect size (Vargha-Delaney *A*), to cover ordinal level data. The R package orddom calculates A as well as bootstrap confidence intervals.

Rank-Biserial Correlation

An effect size related to the common language effect size is the rank-biserial correlation. This measure was introduced by Cureton as an effect size for the Mann-Whitney U test. That is, there are two groups, and scores for the groups have been converted to ranks. The Kerby simple difference formula computes the rank-biserial correlation from the common language effect size. Letting f be the proportion of pairs favorable to the hypothesis (the common language effect size), and letting u be the proportion of pairs not favorable, the rank-biserial r is the simple difference between the two proportions: r = f - u. In other words, the correlation is the difference between the common language effect size and its complement. For example, if the common language effect size is 60%, then the rank-biserial r equals 60% minus 40%, or r = .20. The Kerby formula is directional, with positive values indicating that the results support the hypothesis.

A non-directional formula for the rank-biserial correlation was provided by Wendt, such that the correlation is always positive. The advantage of the Wendt formula is that it can be computed with information that is readily available in published papers. The formula uses only the test value of U from the Mann-Whitney U test, and the sample sizes of the two groups: r = 1 – (2U)/ (n1 * n2). Note that U is defined here according to the classic definition as the smaller of the two U values which can be computed from the data. This ensures that 2*U < n1*n2, as n1*n2 is the maximum value of the U statistics. An example can illustrate the use of the two formulas. Consider a health study of twenty older adults, with ten in the treatment group and ten in the control group; hence, there are ten times ten or 100 pairs. The health program uses diet, exercise, and supplements to improve memory, and memory is measured by a standardized test. A Mann-Whitney U test shows that the adult in the treatment group had the better memory in 70 of the 100 pairs, and the poorer memory in 30 pairs. The Mann-Whitney U is the smaller of 70 and 30, so U = 30. The correlation between memory and treatment performance by the Kerby simple difference formula is r = (70/100) - (30/100) = 0.40. The correlation by the Wendt formula is r = 1 - (2*30) / (10*10) = 0.40.

Effect Size for Ordinal Data

Cliff's delta or d was originally developed by Norman Cliff for use with ordinal data. In short, d is a measure of how often one the values in one distribution are larger than the values in a second distribution. Crucially, it does not require any assumptions about the shape or spread of the two distributions.

The sample estimate d is given by:

$$d = \frac{\#(x_i > x_j) - \#(x_i < x_j)}{mn}$$

where the two distributions are of size n and m with items xi and xj, respectively, and # is defined as the number of times.

d is linearly related to the Mann-Whitney U statistic, however it captures the direction of the difference in its sign. Given the Mann-Whitney U, d is:

$$d = \frac{2U}{mn} - 1$$

The R package orddom calculates d as well as bootstrap confidence intervals.

Confidence Intervals by Means of Noncentrality Parameters

Confidence intervals of standardized effect sizes, especially Cohen's d and f^2, rely on the calculation of confidence intervals of noncentrality parameters (ncp). A common approach to construct the confidence interval of ncp is to find the critical ncp values to fit the observed statistic to tail quantiles $\alpha/2$ and $(1 - \alpha/2)$. The SAS and R-package MBESS provides functions to find critical values of ncp.

t-test for Mean Difference of Single Group or Two Related Groups

For a single group, M denotes the sample mean, μ the population mean, SD the sample's standard deviation, σ the population›s standard deviation, and n is the sample size of the group. The t value is used to test the hypothesis on the difference between the mean and a baseline $\mu_{baseline}$. Usually, $\mu_{baseline}$ is zero. In the case of two related groups, the single group is constructed by the differences in pair of samples, while SD and σ denote the sample›s and population›s standard deviations of differences rather than within original two groups.

$$t := \frac{M}{SE} = \frac{M}{SD/\sqrt{n}} = \frac{\sqrt{n}\frac{M-\mu}{\sigma} + \sqrt{n}\frac{\mu-\mu_{\text{baseline}}}{\sigma}}{\frac{SD}{\sigma}}$$

$$ncp = \sqrt{n}\frac{\mu - \mu_{\text{baseline}}}{\sigma}$$

and Cohen's

$$d := \frac{M - \mu_{\text{baseline}}}{SD}$$

is the point estimate of

$$\frac{\mu - \mu_{\text{baseline}}}{\sigma}.$$

So,

$$\tilde{d} = \frac{ncp}{\sqrt{n}}.$$

t-Test for Mean Difference Between Two Independent Groups

n_1 or n_2 are the respective sample sizes.

$$t := \frac{M_1 - M_2}{SD_{\text{within}}/\sqrt{\frac{n_1 n_2}{n_1+n_2}}},$$

wherein

$$SD_{\text{within}} := \sqrt{\frac{SS_{\text{within}}}{df_{\text{within}}}} = \sqrt{\frac{(n_1-1)SD_1^2 + (n_2-1)SD_2^2}{n_1+n_2-2}}.$$

$$ncp = \sqrt{\frac{n_1 n_2}{n_1+n_2}}\frac{\mu_1 - \mu_2}{\sigma}$$

and Cohen's

$$d := \frac{M_1 - M_2}{SD_{\text{within}}} \text{ is the point estimate of } \frac{\mu_1 - \mu_2}{\sigma}.$$

So,

$$\tilde{d} = \frac{ncp}{\sqrt{\frac{n_1 n_2}{n_1+n_2}}}.$$

One-way ANOVA Test for Mean Difference across Multiple Independent Groups

One-way ANOVA test applies noncentral F distribution. While with a given population standard deviation σ, the same test question applies noncentral chi-squared distribution.

$$F := \frac{\frac{SS_{\text{between}}}{\sigma^2}/df_{\text{between}}}{\frac{SS_{\text{within}}}{\sigma^2}/df_{\text{within}}}$$

For each j-th sample within i-th group $X_{i,j}$, denote

$$M_i\left(X_{i,j}\right) := \frac{\sum_{w=1}^{n_i} X_{i,w}}{n_i};\ \mu_i\left(X_{i,j}\right) := \mu_i.$$

While,

$$\begin{aligned} SS_{\text{between}}/\sigma^2 &= \frac{SS(M_i(X_{i,j}); i=1,2,\ldots,K,\ j=1,2,\ldots,n_i)}{\sigma^2} \\ &= SS\left(\frac{M_i(X_{i,j}-\mu_i)}{\sigma} + \frac{\mu_i}{\sigma}; i=1,2,\ldots,K,\ j=1,2,\ldots,n_i\right) \\ &\sim \chi^2\left(df = K-1,\ ncp = SS\left(\frac{\mu_i(X_{i,j})}{\sigma}; i=1,2,\ldots,K,\ j=1,2,\ldots,n_i\right)\right) \end{aligned}$$

So, both ncp(s) of F and χ^2 equate

$$SS\left(\mu_i(X_{i,j})/\sigma; i=1,2,\ldots,K,\ j=1,2,\ldots,n_i\right).$$

In case of $n := n_1 = n_2 = \cdots = n_K$ for K independent groups of same size, the total sample size is N := n·K.

$$\text{Cohens } \tilde{f}^2 := \frac{SS(\mu_1,\mu_2,\ldots,\mu_K)}{K\cdot\sigma^2} = \frac{SS\left(\mu_i\left(X_{i,j}\right)/\sigma; i=1,2,\ldots,K,\ j=1,2,\ldots,n_i\right)}{n\cdot K} = \frac{ncp}{n\cdot K} = \frac{ncp}{N}.$$

The t-test for a pair of independent groups is a special case of one-way ANOVA. Note that the noncentrality parameter ncp_F of F is not comparable to the noncentrality parameter ncp_t of the corresponding t.

Actually, $ncp_F = ncp_t^2$, and $\tilde{f} = \left|\frac{\tilde{d}}{2}\right|$.

Effect Sizes Descriptors

Whether an effect size should be interpreted small, medium, or large depends on its substantive context and its operational definition. Cohen's conventional criteria small, medium, or big are near ubiquitous across many fields, although Cohen cautioned:

"The terms 'small,' 'medium,' and 'large' are relative, not only to each other, but to the area of behavioral science or even more particularly to the specific content and research method being employed in any given investigation....In the face of this relativity, there is a certain risk inherent in offering conventional operational definitions for these terms for use in power analysis in as diverse a field of inquiry as behavioral science. This risk is nevertheless accepted in the belief that more is to be gained than lost by supplying a common conventional frame of reference which is recommended for use only when no better basis for estimating the ES index is available." (p. 25)

In the two sample layout, Sawilowsky concluded "Based on current research findings in the applied literature, it seems appropriate to revise the rules of thumb for effect sizes," keeping in mind Cohen's cautions, and expanded the descriptions to include very small, very large, and huge. The same de facto standards could be developed for other layouts.

Lenth noted for a "medium" effect size, "you'll choose the same n regardless of the accuracy or reliability of your instrument, or the narrowness or diversity of your subjects. Clearly, important considerations are being ignored here. Researchers should interpret the substantive significance of their results by grounding them in a meaningful context or by quantifying their contribution to knowledge, and Cohen's effect size descriptions can be helpful as a starting point." Similarly, a U.S. Dept of Education sponsored report said "The widespread indiscriminate use of Cohen's generic small, medium, and large effect size values to characterize effect sizes in domains to which his normative values do not apply is thus likewise inappropriate and misleading."

They suggested that "appropriate norms are those based on distributions of effect sizes for comparable outcome measures from comparable interventions targeted on comparable samples." Thus if a study in a field where most interventions are tiny yielded a small effect (by Cohen's criteria), these new criteria would call it "large".

Statistical Power

The power or sensitivity of a binary hypothesis test is the probability that the test correctly rejects the null hypothesis (H_0) when the alternative hypothesis (H_1) is true. It can be equivalently thought of as the probability of accepting the alternative hypothesis (H_1) when it is true—that is, the ability of a test to detect an effect, if the effect actually exists. That is,

$$\text{power} = \mathbb{P}\big(\text{reject} H_0 \big| H_1 \text{ is true}\big)$$

The power of a test sometimes, less formally, refers to the probability of rejecting the null when it is not correct, though this is not the formal definition stated above. The power is in general a function of the possible distributions, often determined by a parameter, under the alternative hypothesis. As the power increases, there are decreasing chances of a Type II error (false negative), which are also referred to as the false negative rate (β) since the power is equal to $1-\beta$, again, under the alternative hypothesis. A similar concept is Type I error, also referred to as the "false positive rate" or the level of a test under the null hypothesis.

Power analysis can be used to calculate the minimum sample size required so that one can be reasonably likely to detect an effect of a given size. For example: "how many times do I need to toss a coin to conclude it is rigged?" Power analysis can also be used to calculate the minimum effect size that is likely to be detected in a study using a given sample size. In addition, the concept of power is used to make comparisons between different statistical testing procedures: for example, between a parametric and a nonparametric test of the same hypothesis.

Background

Statistical tests use data from samples to assess, or make inferences about, a statistical population. In the concrete setting of a two-sample comparison, the goal is to assess whether the mean values of some attribute obtained for individuals in two sub-populations differ. For example, to test the null hypothesis that the mean scores of men and women on a test do not differ, samples of men and women are drawn, the test is administered to them, and the mean score of one group is compared to that of the other group using a statistical test such as the two-sample z-test. The power of the test is the probability that the test will find a statistically significant difference between men and women, as a function of the size of the true difference between those two populations.

Factors Influencing Power

Statistical power may depend on a number of factors. Some factors may be particular to a specific testing situation, but at a minimum, power nearly always depends on the following three factors:

- the statistical significance criterion used in the test
- the magnitude of the effect of interest in the population
- the sample size used to detect the effect

A significance criterion is a statement of how unlikely a positive result must be, if the null hypothesis of no effect is true, for the null hypothesis to be rejected. The most commonly used criteria are probabilities of 0.05 (5%, 1 in 20), 0.01 (1%, 1 in 100), and 0.001 (0.1%, 1 in 1000). If the criterion is 0.05, the probability of the data implying an effect at least as large as the observed effect when the null hypothesis is true must be less than 0.05, for the null hypothesis of no effect to be rejected. One easy way to increase the power of a test is to carry out a less conservative test by using a larger significance criterion, for example 0.10 instead of 0.05. This increases the chance of rejecting the null hypothesis (i.e. obtaining a statistically significant result) when the null hypothesis is false, that is, reduces the risk of a Type II error (false negative regarding whether an effect exists). But it also increases the risk of obtaining a statistically significant result (i.e. rejecting the null hypothesis) when the null hypothesis is not false; that is, it increases the risk of a Type I error (false positive).

The magnitude of the effect of interest in the population can be quantified in terms of an effect size, where there is greater power to detect larger effects. An effect size can be a direct estimate of the quantity of interest, or it can be a standardized measure that also accounts for the variability in the population. For example, in an analysis comparing outcomes in a treated and control population, the difference of outcome means Y − X would be a direct measure of the effect size, whereas (Y − X)/σ where σ is the common standard deviation of the outcomes in the treated and control groups, would be a standardized effect size. If constructed appropriately, a standardized effect size, along with the sample size, will completely determine the power. An unstandardized (direct) effect size will rarely be sufficient to determine the power, as it does not contain information about the variability in the measurements.

The sample size determines the amount of sampling error inherent in a test result. Other things being equal, effects are harder to detect in smaller samples. Increasing sample size is often the easiest way to boost the statistical power of a test. How increased sample size translates to higher power is a measure of the efficiency of the test—for example, the sample size required for a given power.

The precision with which the data are measured also influences statistical power. Consequently, power can often be improved by reducing the measurement error in the data. A related concept is to improve the "reliability" of the measure being assessed (as in psychometric reliability).

The design of an experiment or observational study often influences the power. For example, in a two-sample testing situation with a given total sample size n, it is optimal to have equal numbers of observations from the two populations being compared (as long as the variances in the two populations are the same). In regression analysis and Analysis of Variance, there are extensive theories and practical strategies for improving the power based on optimally setting the values of the independent variables in the model.

Interpretation

Although there are no formal standards for power (sometimes referred to as π), most researchers assess the power of their tests using π=0.80 as a standard for adequacy. This convention implies a four-to-one trade off between β-risk and α-risk. (β is the probability of a Type II error; α is the probability of a Type I error, 0.2 and 0.05 are conventional values for β and α). However, there will be times when this 4-to-1 weighting is inappropriate. In medicine, for example, tests are often designed in such a way that no false negatives (Type II errors) will be produced. But this inevitably raises the risk of obtaining a false positive (a Type I error). The rationale is that it is better to tell a healthy patient "we may have found something—let›s test further," than to tell a diseased patient "all is well."

Power analysis is appropriate when the concern is with the correct rejection, or not, of a null hypothesis. In many contexts, the issue is less about determining if there is or is not a difference but rather with getting a more refined estimate of the population effect size. For example, if we were expecting a population correlation between intelligence and job performance of around 0.50, a sample size of 20 will give us approximately 80% power (alpha = 0.05, two-tail) to reject the null hypothesis of zero correlation. However, in doing this study we are probably more interested in knowing whether the correlation is 0.30 or 0.60 or 0.50. In this context we would need a much larger sample size in order to reduce the confidence interval of our estimate to a range that is acceptable for our purposes. Techniques similar to those employed in a traditional power analysis can be used to determine the sample size required for the width of a confidence interval to be less than a given value.

Many statistical analyses involve the estimation of several unknown quantities. In simple cases, all but one of these quantities is a nuisance parameter. In this setting, the only relevant power pertains to the single quantity that will undergo formal statistical inference. In some settings, particularly if the goals are more «exploratory», there may be a number of quantities of interest in the analysis. For example, in a multiple regression analysis we may include several covariates of potential interest. In situations such as this where several hypotheses are under consideration, it is common that the powers associated with the different hypotheses differ. For instance, in multiple regression analysis, the power for detecting an effect of a given size is related to the variance of the covariate. Since different covariates will have different variances, their powers will differ as well.

Any statistical analysis involving multiple hypotheses is subject to inflation of the type I error rate if appropriate measures are not taken. Such measures typically involve applying a higher threshold of stringency to reject a hypothesis in order to compensate for the multiple comparisons being made (e.g. as in the Bonferroni method). In this situation, the power analysis should reflect the multiple testing approach to be used. Thus, for example, a given study may be well powered to detect a certain effect size when only one test is to be made, but the same effect size may have much lower power if several tests are to be performed.

It is also important to consider the statistical power of a hypothesis test when interpreting its results. A test's power is the probability of correctly rejecting the null hypothesis when it is false; a test's power is influenced by the choice of significance level for the test, the size of the effect being measured, and the amount of data available. A hypothesis test may fail to reject the null, for example, if a true difference exists between two populations being compared by a t-test but the effect is small and the sample size is too small to distinguish the effect from random chance. Many clinical trials, for instance, have low statistical power to detect differences in adverse effects of treatments, since such effects may be rare and the number of affected patients small.

A Priori vs. post hoc Analysis

Power analysis can either be done before (a priori or prospective power analysis) or after (post hoc or retrospective power analysis) data are collected. A priori power analysis is conducted prior to the research study, and is typically used in estimating sufficient sample sizes to achieve adequate power. Post-hoc power analysis is conducted after a study has been completed, and uses the obtained sample size and effect size to determine what the power was in the study, assuming the effect size in the sample is equal to the effect size in the population. Whereas the utility of prospective power analysis in experimental design is universally accepted, the usefulness of retrospective techniques is controversial. Falling for the temptation to use the statistical analysis of the collected data to estimate the power will result in uninformative and misleading values. In particular, it has been shown that post-hoc power in its simplest form is a one-to-one function of the p-value attained. This has been extended to show that all post-hoc power analyses suffer from what is called the "power approach paradox" (PAP), in which a study with a null result is thought to show MORE evidence that the null hypothesis is actually true when the p-value is smaller, since the apparent power to detect an actual effect would be higher. In fact, a smaller p-value is properly understood to make the null hypothesis LESS likely to be true.

Application

Funding agencies, ethics boards and research review panels frequently request that a researcher perform a power analysis, for example to determine the minimum number of animal test subjects needed for an experiment to be informative. In frequentist statistics, an underpowered study is unlikely to allow one to choose between hypotheses at the desired significance level. In Bayesian statistics, hypothesis testing of the type used in classical power analysis is not done. In the Bayesian framework, one updates his or her prior beliefs using the data obtained in a given study. In principle, a study that would be deemed underpowered from the perspective of hypothesis testing could still be used in such an updating process. However, power remains a useful measure of how much a given experiment size can be expected to refine one's beliefs. A study with low power is unlikely to lead to a large change in beliefs.

Sample Size Determination

Sample size determination is the act of choosing the number of observations or replicates to include in a statistical sample. The sample size is an important feature of any empirical study in which the goal is to make inferences about a population from a sample. In practice, the sample size used in a study is determined based on the expense of data collection, and the need to have sufficient statistical power. In complicated studies there may be several different sample sizes involved in the study: for example, in a stratified survey there would be different sample sizes for each stratum. In a census, data are collected on the entire population, hence the sample size is equal to the population size. In experimental design, where a study may be divided into different treatment groups, there may be different sample sizes for each group.

Sample sizes may be chosen in several different ways:

- experience - For example, include those items readily available or convenient to collect. A choice of small sample sizes, though sometimes necessary, can result in wide confidence intervals or risks of errors in statistical hypothesis testing.
- using a target variance for an estimate to be derived from the sample eventually obtained
- using a target for the power of a statistical test to be applied once the sample is collected.
- using a confidence level determines how accurate a result will turn out with lower chances of error.

Larger sample sizes generally lead to increased precision when estimating unknown parameters. For example, if we wish to know the proportion of a certain species of fish that is infected with a pathogen, we would generally have a more precise estimate of this proportion if we sampled and examined 200 rather than 100 fish. Several fundamental facts of mathematical statistics describe this phenomenon, including the law of large numbers and the central limit theorem.

In some situations, the increase in precision for larger sample sizes is minimal, or even non-existent. This can result from the presence of systematic errors or strong dependence in the data, or if the data follows a heavy-tailed distribution.

Sample sizes are judged based on the quality of the resulting estimates. For example, if a proportion is being estimated, one may wish to have the 95% confidence interval be less than 0.06 units wide. Alternatively, sample size may be assessed based on the power of a hypothesis test. For example, if we are comparing the support for a certain political candidate among women with the support for that candidate among men, we may wish to have 80% power to detect a difference in the support levels of 0.04 units.

Estimation

A relatively simple situation is estimation of a proportion. For example, we may wish to estimate the proportion of residents in a community who are at least 65 years old.

The estimator of a proportion is $\hat{p} = X/n$, where X is the number of 'positive' observations (e.g. the number of people out of the n sampled people who are at least 65 years old). When the observations are independent, this estimator has a (scaled) binomial distribution (and is also the sample mean of data from a Bernoulli distribution). The maximum variance of this distribution is 0.25/n, which occurs when the true parameter is p = 0.5. In practice, since p is unknown, the maximum variance is often used for sample size assessments.

For sufficiently large n, the distribution of $\hat{p}$ will be closely approximated by a normal distribution. Using this approximation, it can be shown that around 95% of this distribution's probability lies within 2 standard deviations of the mean. Using the Wald method for the binomial distribution, an interval of the form

$$\left(\hat{p} - 2\sqrt{\frac{0.25}{n}}, \hat{p} + 2\sqrt{\frac{0.25}{n}}\right)$$

will form a 95% confidence interval for the true proportion. If this interval needs to be no more than W units wide, the equation

$$4\sqrt{\frac{0.25}{n}} = W$$

can be solved for n, yielding $n = 4/W^2 = 1/B^2$ where B is the error bound on the estimate, i.e., the estimate is usually given as within ± B. So, for B = 10% one requires n = 100, for B = 5% one needs n = 400, for B = 3% the requirement approximates to n = 1000, while for B = 1% a sample size of n = 10000 is required. These numbers are quoted often in news reports of opinion polls and other sample surveys.

Means

A proportion is a special case of a mean. When estimating the population mean using an independent and identically distributed (iid) sample of size n, where each data value has variance σ^2, the standard error of the sample mean is:

$$\frac{\sigma}{\sqrt{n}}.$$

This expression describes quantitatively how the estimate becomes more precise as the sample size increases. Using the central limit theorem to justify approximating the sample mean with a normal distribution yields an approximate 95% confidence interval of the form

$$\left(\bar{x} - \frac{2\sigma}{\sqrt{n}}, \bar{x} + \frac{2\sigma}{\sqrt{n}}\right).$$

If we wish to have a confidence interval that is W units in width, we would solve

$$\frac{4\sigma}{\sqrt{n}} = W$$

for n, yielding the sample size $n = 16\sigma^2/W^2$.

For example, if we are interested in estimating the amount by which a drug lowers a subject's blood pressure with a confidence interval that is six units wide, and we know that the standard deviation of blood pressure in the population is 15, then the required sample size is 100.

Required Sample Sizes for Hypothesis Tests

A common problem faced by statisticians is calculating the sample size required to yield a certain power for a test, given a predetermined Type I error rate α. As follows, this can be estimated by pre-determined tables for certain values, by Mead›s resource equation, or, more generally, by the cumulative distribution function:

Tables

The table shown on the right can be used in a two-sample t-test to estimate the sample sizes of an experimental group and a control group that are of equal size, that is, the total number of individuals in the trial is twice that of the number given, and the desired significance level is 0.05. The parameters used are:

- The desired statistical power of the trial, shown in column to the left.
- Cohen's d (=effect size), which is the expected difference between the means of the target values between the experimental group and the control group, divided by the expected standard deviation.

Power	Cohen's d		
	0.2	0.5	0.8
0.25	84	14	6
0.50	193	32	13
0.60	246	40	16
0.70	310	50	20
0.80	393	64	26
0.90	526	85	34
0.95	651	105	42
0.99	920	148	58

Mead's Resource Equation

Mead's resource equation is often used for estimating sample sizes of laboratory animals, as well as in many other laboratory experiments. It may not be as accurate as using other methods in estimating sample size, but gives a hint of what is the appropriate sample size where parameters such as expected standard deviations or expected differences in values between groups are unknown or very hard to estimate.

All the parameters in the equation are in fact the degrees of freedom of the number of their concepts, and hence, their numbers are subtracted by 1 before insertion into the equation.

The equation is:

$$E = N - B - T,$$

where:

- N is the total number of individuals or units in the study (minus 1)
- B is the blocking component, representing environmental effects allowed for in the design (minus 1)
- T is the treatment component, corresponding to the number of treatment groups (including control group) being used, or the number of questions being asked (minus 1)
- E is the degrees of freedom of the error component, and should be somewhere between 10 and 20.

For example, if a study using laboratory animals is planned with four treatment groups (T=3), with eight animals per group, making 32 animals total (N=31), without any further stratification (B=0), then E would equal 28, which is above the cutoff of 20, indicating that sample size may be a bit too large, and six animals per group might be more appropriate.

Cumulative Distribution Function

Let X_i, i = 1, 2, ..., n be independent observations taken from a normal distribution with unknown mean μ and known variance σ^2. Let us consider two hypotheses, a null hypothesis:

$$H_0 : \mu = 0$$

and an alternative hypothesis:

$$H_a : \mu = \mu^*$$

for some 'smallest significant difference' $\mu^* > 0$. This is the smallest value for which we care about observing a difference. Now, if we wish to (1) reject H_0 with a probability of at least 1-β when H_a is true (i.e. a power of 1-β), and (2) reject H_0 with probability α when H_0 is true, then we need the following:

If z_α is the upper α percentage point of the standard normal distribution, then

$$\Pr(\bar{x} > z_\alpha \sigma/\sqrt{n} | H_0 \text{ true}) = \alpha$$

and so

'Reject H_0 if our sample average ($\bar{x}$) is more than $z_\alpha \sigma/\sqrt{n}$'

is a decision rule which satisfies (2). (Note, this is a 1-tailed test)

Now we wish for this to happen with a probability at least 1-β when H_a is true. In this case, our sample average will come from a Normal distribution with mean μ^*. Therefore, we require

$$\Pr(\bar{x} > z_\alpha \sigma/\sqrt{n} | H_a \text{ true}) \geq 1 - \beta$$

Through careful manipulation, this can be shown to happen when

$$n \geq \left(\frac{z_\alpha + \Phi^{-1}(1-\beta)}{\mu^*/\sigma} \right)^2$$

where Φ is the normal cumulative distribution function.

Stratified Sample Size

With more complicated sampling techniques, such as stratified sampling, the sample can often be split up into sub-samples. Typically, if there are H such sub-samples (from H different strata) then each of them will have a sample size n_h, h = 1, 2, ..., H. These n_h must conform to the rule that $n_1 + n_2 + ... + n_H = n$ (i.e. that the total sample size is given by the sum of the sub-sample sizes). Selecting these n_h optimally can be done in various ways, using (for example) Neyman's optimal allocation.

There are many reasons to use stratified sampling: to decrease variances of sample estimates, to use partly non-random methods, or to study strata individually. A useful, partly non-random method would be to sample individuals where easily accessible, but, where not, sample clusters to save travel costs.

In general, for H strata, a weighted sample mean is

$$\bar{x}_w = \sum_{h=1}^{H} W_h \bar{x}_h,$$

with

$$\mathrm{Var}(\bar{x}_w) = \sum_{h=1}^{H} W_h^2 \ \mathrm{Var}(\bar{x}_h).$$

The weights, W(h), frequently, but not always, represent the proportions of the population elements in the strata, and $W_h = N_h/N$. For a fixed sample size, that is $N = \sum N_h$,

$$\mathrm{Var}(\bar{x}_w) = \sum_{h=1}^{H} W_h^2 \, Var_h \left(\frac{1}{n_h} - \frac{1}{N_h}\right),$$

which can be made a minimum if the sampling rate within each stratum is made proportional to the standard deviation within each stratum: $n_h/N_h = kS_h$, where $S_h = \sqrt{Var_h}$ and k is a constant such that $\sum n_h = n$.

An "optimum allocation" is reached when the sampling rates within the strata are made directly proportional to the standard deviations within the strata and inversely proportional to the square root of the sampling cost per element within the strata, C_h:

$$\frac{n_h}{N_h} = \frac{KS_h}{\sqrt{C_h}},$$

where k is a constant such that $\sum n_h = n$, or, more generally, when

$$n_h = \frac{K' W_h S_h}{\sqrt{C_h}}.$$

Qualitative Research

Sample size determination in qualitative studies takes a different approach. It is generally a subjective judgment, taken as the research proceeds. One approach is to continue to include further participants or material until saturation is reached. The number needed to reach saturation has been investigated empirically.

There is a paucity of reliable guidance on estimating sample sizes before starting the research, with a range of suggestions given. A tool akin to a quantitative power calculation, based on the negative binomial distribution, has been suggested for thematic analysis.

Missing Data

In statistics, missing data, or missing values, occur when no data value is stored for the variable in an observation. Missing data are a common occurrence and can have a significant effect on the conclusions that can be drawn from the data. Missing data can occur because of nonresponse: no information is provided for one or more items or for a whole unit ("subject"). Some items are more likely to generate a nonresponse than others: for example items about private subjects such as income. Attrition ("Dropout") is a type of missingness that can occur in longitudinal studies - for instance studying development where a measurement is repeated after a certain period of time. Missingness occurs when participants drop out before the test ends and one or more measurements are missing. Data often are missing in research in economics, sociology, and political science because governments choose not to, or fail to, report critical statistics. Sometimes missing values are caused by the researcher—for example, when data collection is done improperly or mistakes are made in data entry. These forms of missingness take different types, with different impacts on the validity of conclusions from research: Missing completely at random, missing at random, and missing not at random.

Types of Missing Data

Understanding the reasons why data are missing is important to correctly handle the remaining data. If values are missing completely at random, the data sample is likely still representative of the population. But if the values are missing systematically, analysis may be biased. For example, in a study of the relation between IQ and income, if participants with an above-average IQ tend to skip the question 'What is your salary?' , analyses that do not take into account this missing at random (MAR pattern (see below)) may falsely fail to find a positive association between IQ and salary. Because of these problems, methodologists routinely advise researchers to design studies to minimize the occurrence of missing values. Graphical models can be used to describe the missing data mechanism in detail.

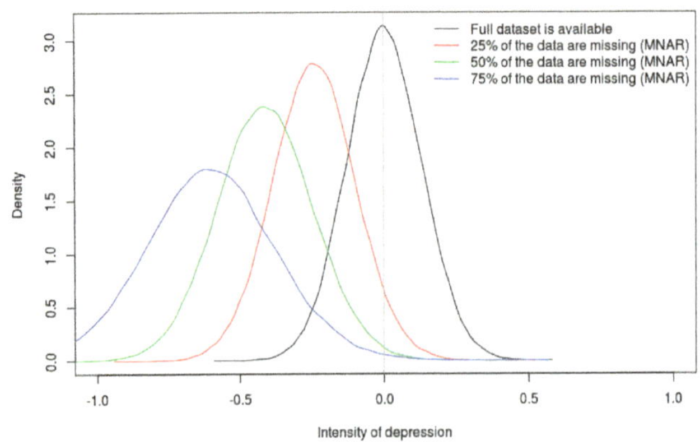

The graph shows the probability distributions of the estimations of the expected intensity of depression in the population. The number of cases is 60. Let the true population be a standardized normal distribution and the non-response probability be a logistic function of the intensity of depression. The conclusion is: The more data is missing (MNAR), the more biased are the estimations. We underestimate the intensity of depression in the population.

Missing Completely at Random

Values in a data set are missing completely at random (MCAR) if the events that lead to any particular data-item being missing are independent both of observable variables and of unobservable parameters of interest, and occur entirely at random. When data are MCAR, the analysis performed on the data is unbiased; however, data are rarely MCAR.

In the case of MCAR, the missingness of data is unrelated to any study variable: thus, the participants with completely observed data are in effect a random sample of all the participants assigned a particular intervention. With MCAR, the random assignment of treatments is assumed to be preserved, but that is usually an unrealistically strong assumption in practice.

Missing at Random

Missing at random (MAR) occurs when the missingness is not random, but where missingness can be fully accounted for by variables where there is complete information. MAR is an assumption that is impossible to verify statistically, we must rely on its substantive reasonableness. An example is that males are less likely to fill in a depression survey but this has nothing to do with their level of depression, after accounting for maleness. These data can still induce parameter bias in analyses due to the contingent emptyness of cells (male, very high depression may have zero entries).

Missing not at Random

Missing not at random (MNAR) (also known as nonignorable nonresponse) is data that is neither MAR nor MCAR (i.e. the value of the variable that's missing is related to the reason it's missing). To extend the previous example, this would occur if men failed to fill in a depression survey because of their level of depression.

Techniques of Dealing with Missing Data

Missing data reduce the representativeness of the sample and can therefore distort inferences about the population. If it is possible try to think about how to prevent data from missingness before the actual data gathering takes place. For example, in computer questionnaires it is often not possible to skip a question. A question has to be answered, otherwise one cannot continue to the next. So missing values due to the participant are eliminated by this type of questionnaire, though this method may not be permitted by an ethics board overseeing the research. And in survey research, it is common to make multiple efforts to contact each individual in the sample, often sending letters to attempt to persuade those who have decided not to participate to change their minds.[:161–187] However, such techniques can either help or hurt in terms of reducing the negative inferential effects of missing data, because the kind of people who are willing to be persuaded to participate after initially refusing or not being home are likely to be significantly different from the kinds of people who will still refuse or remain unreachable after additional effort.[:188–198]

In situations where missing data are likely to occur, the researcher is often advised to plan to use methods of data analysis methods that are robust to missingness. An analysis is robust when we are confident that mild to moderate violations of the technique's key assumptions will produce little or no bias, or distortion in the conclusions drawn about the population.

Imputation

If it is known that the data analysis technique which is to be used is not content robust, it is good to consider imputing the missing data. This can be done in several ways. Recommended is to use multiple imputations. Rubin (1987) argued that even a small number (5 or fewer) of repeated imputations enormously improves the quality of estimation.

For many practical purposes, 2 or 3 imputations capture most of the relative efficiency that could be captured with a larger number of imputations. However, a too-small number of imputations can lead to a substantial loss of statistical power, and some scholars now recommend 20 to 100 or more. Any multiply-imputed data analysis must be repeated for each of the imputed data sets and, in some cases, the relevant statistics must be combined in a relatively complicated way.

Examples of imputations are listed below.

Partial Imputation

The expectation-maximization algorithm is an approach in which values of the statistics which would be computed if a complete dataset were available are estimated (imputed), taking into account the pattern of missing data. In this approach, values for individual missing data-items are not usually imputed.

Partial Deletion

Methods which involve reducing the data available to a dataset having no missing values include:

- Listwise deletion/casewise deletion
- Pairwise deletion

Full Analysis

Methods which take full account of all information available, without the distortion resulting from using imputed values as if they were actually observed:

- The expectation-maximization algorithm
- full information maximum likelihood estimation

Interpolation

In the mathematical field of numerical analysis, interpolation is a method of constructing new data points within the range of a discrete set of known data points.

Model-Based Techniques

Model based techniques, often using graphs, offer additional tools for testing missing data types (MCAR, MAR, MNAR) and for estimating parameters under missing data conditions. For example, a test for refuting MAR/MCAR reads as follows:

For any three variables X,Y, and Z where Z is fully observed and X and Y partially observed, the data should satisfy: $X \perp\!\!\!\perp R_y|(R_x, Z)$.

In words, the observed portion of X should be independent on the missingness status of Y, conditional on every value of Z. Failure to satisfy this condition indicates that the problem belongs to the MNAR category.

(Remark: These tests are necessary for variable-based MAR which is a slight variation of event-based MAR.)

When data falls into MNAR category techniques are available for consistently estimating parameters when certain conditions hold in the model. For example, if Y explains the reason for missingness in X and Y itself has missing values, the joint probability distribution of X and Y can still be estimated if the missingness of Y is random. The estimand in this case will be:

$$\begin{aligned} P(X,Y) &= P(X|Y)P(Y) \\ &= P(X|Y, R_x = 0, R_y = 0)P(Y|R_y = 0) \end{aligned}$$

where $R_x = 0$ and $R_y = 0$ denote the observed portions of their respective variables.

Different model structures may yield different estimands and different procedures of estimation whenever consistent estimation is possible. The preceding estimand calls for first estimating $P(X|Y)$ from complete data and multiplying it by P(Y) estimated from cases in which Y is observed regardless of the status of X. Moreover, in order to obtain a consistent estimate it is crucial that the first term be $P(X|Y)$ as opposed to $P(Y|X)$.

In many cases model based techniques permit the model structure to undergo refutation tests. Any model which implies the independence between a partially observed variable X and the missingness indicator of another variable Y (i.e. R(y)), conditional on R(x) can be submitted to the following refutation test: $X \perp\!\!\!\perp R_y|R_x = 0$.

Finally, the estimands that emerge from these techniques are derived in closed form and do not require iterative procedures such as Expectation Maximization that are susceptible to local optima.

A special class of problems appears when the probability of the missingness depends on time. For example, in the trauma databases the probability to loose data about the trauma outcome depends on the day after trauma. In these cases various non-stationary Markov chain models are applied.

SURVEY METHODOLOGY

Sampling

In statistics, quality assurance, and survey methodology, sampling is concerned with the selection of a subset of individuals from within a statistical population to estimate characteristics of the whole population. Two advantages of sampling are that the cost is lower and data collection is faster than measuring the entire population.

Each observation measures one or more properties (such as weight, location, color) of observable bodies distinguished as independent objects or individuals. In survey sampling, weights can be applied to the data to adjust for the sample design, particularly stratified sampling. Results from probability theory and statistical theory are employed to guide the practice. In business and medical research, sampling is widely used for gathering information about a population. Acceptance sampling is used to determine if a production lot of material meets the governing specifications.

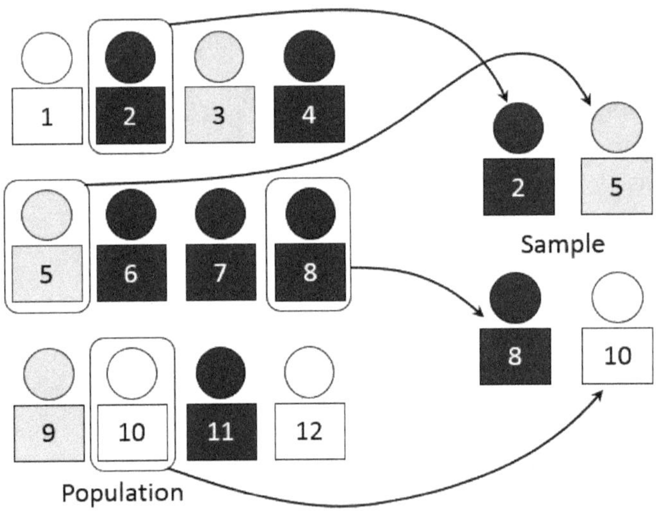

A visual representation of the sampling process.

The sampling process comprises several stages:

- Defining the population of concern
- Specifying a sampling frame, a set of items or events possible to measure
- Specifying a sampling method for selecting items or events from the frame
- Determining the sample size
- Implementing the sampling plan
- Sampling and data collecting

Population Definition

Successful statistical practice is based on focused problem definition. In sampling, this includes defining the population from which our sample is drawn. A population can be defined as including all people or items with the characteristic one wishes to understand. Because there is very rarely enough time or money to gather information from everyone or everything in a population, the goal becomes finding a representative sample (or subset) of that population.

Sometimes what defines a population is obvious. For example, a manufacturer needs to decide whether a batch of material from production is of high enough quality to be released to the customer, or should be sentenced for scrap or rework due to poor quality. In this case, the batch is the population.

Although the population of interest often consists of physical objects, sometimes we need to sample over time, space, or some combination of these dimensions. For instance, an investigation of supermarket staffing could examine checkout line length at various times, or a study on endangered penguins might aim to understand their usage of various hunting grounds over time. For the time dimension, the focus may be on periods or discrete occasions.

In other cases, our 'population' may be even less tangible. For example, Joseph Jagger studied the behavior of roulette wheels at a casino in Monte Carlo, and used this to identify a biased wheel. In this case, the 'population' Jagger wanted to investigate was the overall behavior of the wheel (i.e. the probability distribution of its results over infinitely many trials), while his 'sample' was formed from observed results from that wheel. Similar considerations arise when taking repeated measurements of some physical characteristic such as the electrical conductivity of copper.

This situation often arises when we seek knowledge about the cause system of which the observed population is an outcome. In such cases, sampling theory may treat the observed population as a sample from a larger 'superpopulation'. For example, a researcher might study the success rate of a new 'quit smoking' program on a test group of 100 patients, in order to predict the effects of the program if it were made available nationwide. Here the superpopulation is "everybody in the country, given access to this treatment" – a group which does not yet exist, since the program isn't yet available to all.

Note also that the population from which the sample is drawn may not be the same as the population about which we actually want information. Often there is large but not complete overlap between these two groups due to frame issues etc. (see below). Sometimes they may be entirely separate – for instance, we might study rats in order to get a better understanding of human health, or we might study records from people born in 2008 in order to make predictions about people born in 2009.

Time spent in making the sampled population and population of concern precise is often well spent, because it raises many issues, ambiguities and questions that would otherwise have been overlooked at this stage.

Sampling Frame

In the most straightforward case, such as the sampling of a batch of material from production (acceptance sampling by lots), it would be most desirable to identify and measure every single item in the population and to include any one of them in our sample. However, in the more general case this is not usually possible or practical. There is no way to identify all rats in the set of all rats. Where voting is not compulsory, there is no way to identify which people will actually vote at a forthcoming election (in advance of the election). These imprecise populations are not amenable to sampling in any of the ways below and to which we could apply statistical theory.

As a remedy, we seek a sampling frame which has the property that we can identify every single element and include any in our sample. The most straightforward type of frame is a list of elements of the population (preferably the entire population) with appropriate contact information. For example, in an opinion poll, possible sampling frames include an electoral register and a telephone directory.

A probability sample is a sample in which every unit in the population has a chance (greater than zero) of being selected in the sample, and this probability can be accurately determined. The combination of these traits makes it possible to produce unbiased estimates of population totals, by weighting sampled units according to their probability of selection.

Example: We want to estimate the total income of adults living in a given street. We visit each household in that street, identify all adults living there, and randomly select one adult from each household. (For example, we can allocate each person a random number, generated from a uniform distribution between 0 and 1, and select the person with the highest number in each household). We then interview the selected person and find their income.

People living on their own are certain to be selected, so we simply add their income to our estimate of the total. But a person living in a household of two adults has only a one-in-two chance of selection. To reflect this, when we come to such a household, we would count the selected person's income twice towards the total. (The person who is selected from that household can be loosely viewed as also representing the person who isn't selected.)

In the above example, not everybody has the same probability of selection; what makes it a probability sample is the fact that each person's probability is known. When every element in the population does have the same probability of selection, this is known as an 'equal probability of selection' (EPS) design. Such designs are also referred to as 'self-weighting' because all sampled units are given the same weight.

Probability sampling includes: Simple Random Sampling, Systematic Sampling, Stratified Sampling, Probability Proportional to Size Sampling, and Cluster or Multistage Sampling. These various ways of probability sampling have two things in common:

- Every element has a known nonzero probability of being sampled and
- involves random selection at some point.

Nonprobability Sampling

Nonprobability sampling is any sampling method where some elements of the population have no chance of selection (these are sometimes referred to as 'out of coverage'/'undercovered'), or where the probability of selection can't be accurately determined. It involves the selection of elements based on assumptions regarding the population of interest, which forms the criteria for selection. Hence, because the selection of elements is nonrandom, nonprobability sampling does not allow the estimation of sampling errors. These conditions give rise to exclusion bias, placing limits on how much information a sample can provide about the population. Information about the relationship between sample and population is limited, making it difficult to extrapolate from the sample to the population.

Example: We visit every household in a given street, and interview the first person to answer the door. In any household with more than one occupant, this is a nonprobability sample, because some people are more likely to answer the door (e.g. an unemployed person who spends most of their time at home is more likely to answer than an employed housemate who might be at work when the interviewer calls) and it's not practical to calculate these probabilities.

Nonprobability sampling methods include convenience sampling, quota sampling and purposive sampling. In addition, nonresponse effects may turn any probability design into a nonprobability design if the characteristics of nonresponse are not well understood, since nonresponse effectively modifies each element's probability of being sampled.

Sampling Methods

Within any of the types of frame identified above, a variety of sampling methods can be employed, individually or in combination. Factors commonly influencing the choice between these designs include:

- Nature and quality of the frame
- Availability of auxiliary information about units on the frame
- Accuracy requirements, and the need to measure accuracy
- Whether detailed analysis of the sample is expected
- Cost/operational concerns

Simple Random Sampling

In a simple random sample (SRS) of a given size, all such subsets of the frame are given an equal probability. Furthermore, any given pair of elements has the same chance of selection as any other such pair (and similarly for triples, and so on). This minimizes bias and simplifies analysis of results. In particular, the variance between individual results within the sample is a good indicator of variance in the overall population, which makes it relatively easy to estimate the accuracy of results.

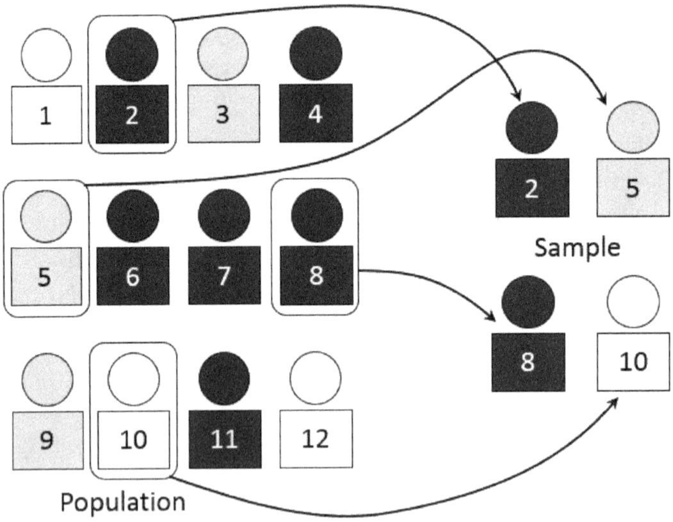

A visual representation of selecting a simple random sample

SRS can be vulnerable to sampling error because the randomness of the selection may result in a sample that doesn't reflect the makeup of the population. For instance, a simple random sample of ten people from a given country will on average produce five men and five women, but any given trial is likely to over represent one sex and underrepresent the other. Systematic and stratified techniques attempt to overcome this problem by "using information about the population" to choose a more "representative" sample.

SRS may also be cumbersome and tedious when sampling from an unusually large target population. In some cases, investigators are interested in "research questions specific" to subgroups of the population. For example, researchers might be interested in examining whether cognitive ability as a predictor of job performance is equally applicable across racial groups. SRS cannot accommodate the needs of researchers in this situation because it does not provide subsamples of the population. "Stratified sampling" addresses this weakness of SRS.

Systematic Sampling

Systematic sampling (also known as interval sampling) relies on arranging the study population according to some ordering scheme and then selecting elements at regular intervals through that ordered list. Systematic sampling involves a random start and then proceeds with the selection of every kth element from then onwards. In this case, k=(population size/sample size). It is important that the starting point is not automatically the first in the list, but is instead randomly chosen from within the first to the kth element in the list. A simple example would be to select every 10th name from the telephone directory (an 'every 10th' sample, also referred to as 'sampling with a skip of 10').

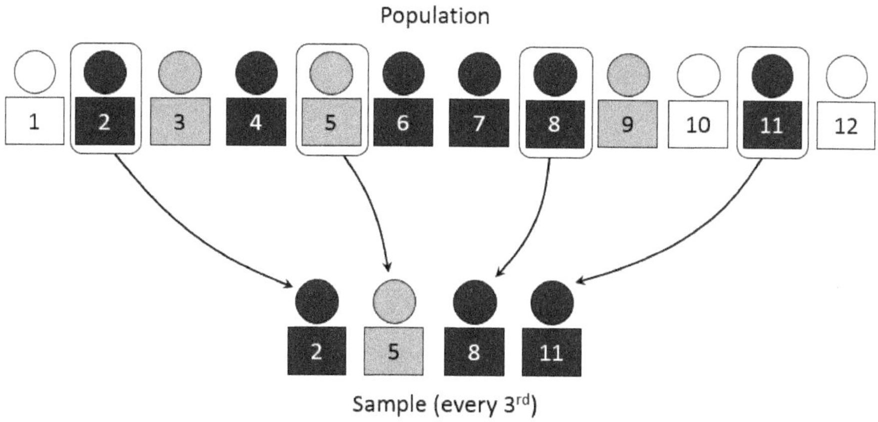

A visual representation of selecting a random sample using the systematic sampling technique

As long as the starting point is randomized, systematic sampling is a type of probability sampling. It is easy to implement and the stratification induced can make it efficient, if the variable by which the list is ordered is correlated with the variable of interest. 'Every 10th' sampling is especially useful for efficient sampling from databases.

For example, suppose we wish to sample people from a long street that starts in a poor area (house No. 1) and ends in an expensive district (house No. 1000). A simple random selection of addresses from this street could easily end up with too many from the high end and too few from the low end (or vice versa), leading to an unrepresentative sample. Selecting (e.g.) every 10th street number along the street ensures that the sample is spread evenly along the length of the street, representing all of these districts. (Note that if we always start at house #1 and end at #991, the sample is slightly biased towards the low end; by randomly selecting the start between #1 and #10, this bias is eliminated.

However, systematic sampling is especially vulnerable to periodicities in the list. If periodicity is present and the period is a multiple or factor of the interval used, the sample is especially likely to be unrepresentative of the overall population, making the scheme less accurate than simple random sampling.

For example, consider a street where the odd-numbered houses are all on the north (expensive) side of the road, and the even-numbered houses are all on the south (cheap) side. Under the sampling scheme given above, it is impossible to get a representative sample; either the houses sampled will all be from the odd-numbered, expensive side, or they will all be from the even-numbered, cheap side, unless the researcher has previous knowledge of this bias and avoids it by a using a skip which ensures jumping between the two sides (any odd-numbered skip).

Another drawback of systematic sampling is that even in scenarios where it is more accurate than SRS, its theoretical properties make it difficult to quantify that accuracy. (In the two examples of systematic sampling that are given above, much of the potential sampling error is due to variation between neighboring houses – but because this method never selects two neighboring houses, the sample will not give us any information on that variation.)

As described above, systematic sampling is an EPS method, because all elements have the same probability of selection (in the example given, one in ten). It is not 'simple random sampling' because different subsets of the same size have different selection probabilities – e.g. the set {4,14,24,...,994} has a one-in-ten probability of selection, but the set {4,13,24,34,...} has zero probability of selection.

Systematic sampling can also be adapted to a non-EPS approach; for an example, see discussion of PPS samples below.

Stratified Sampling

When the population embraces a number of distinct categories, the frame can be organized by these categories into separate "strata." Each stratum is then sampled as an independent sub-population, out of which individual elements can be randomly selected. There are several potential benefits to stratified sampling.

First, dividing the population into distinct, independent strata can enable researchers to draw inferences about specific subgroups that may be lost in a more generalized random sample.

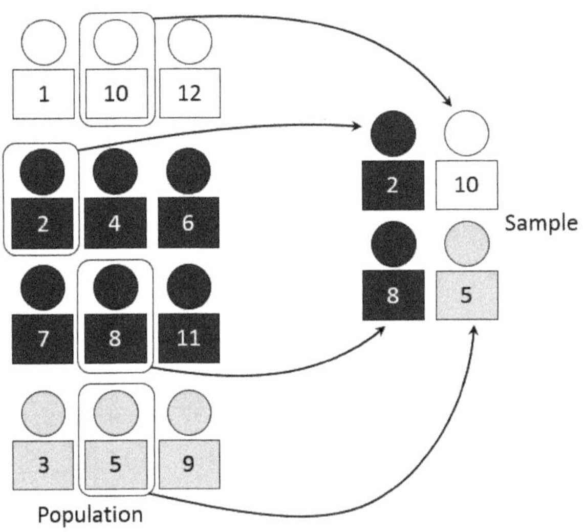

A visual representation of selecting a random sample using the stratified sampling technique

Second, utilizing a stratified sampling method can lead to more efficient statistical estimates (provided that strata are selected based upon relevance to the criterion in question, instead of availability of the samples). Even if a stratified sampling approach does not lead to increased statistical efficiency, such a tactic will not result in less efficiency than would simple random sampling, provided that each stratum is proportional to the group's size in the population.

Third, it is sometimes the case that data are more readily available for individual, pre-existing strata within a population than for the overall population; in such cases, using a stratified sampling approach may be more convenient than aggregating data across groups (though this may potentially be at odds with the previously noted importance of utilizing criterion-relevant strata).

Finally, since each stratum is treated as an independent population, different sampling approaches can be applied to different strata, potentially enabling researchers to use the approach best suited (or most cost-effective) for each identified subgroup within the population.

There are, however, some potential drawbacks to using stratified sampling. First, identifying strata and implementing such an approach can increase the cost and complexity of sample selection, as well as leading to increased complexity of population estimates. Second, when examining multiple criteria, stratifying variables may be related to some, but not to others, further complicating the design, and potentially reducing the utility of the strata. Finally, in some cases (such as designs with a large number of strata, or those with a specified minimum sample size per group), stratified sampling can potentially require a larger sample than would other methods (although in most cases, the required sample size would be no larger than would be required for simple random sampling.

A stratified sampling approach is most effective when three conditions are met

- Variability within strata are minimized
- Variability between strata are maximized
- The variables upon which the population is stratified are strongly correlated with the desired dependent variable.

Advantages over other sampling methods

- Focuses on important subpopulations and ignores irrelevant ones.
- Allows use of different sampling techniques for different subpopulations.
- Improves the accuracy/efficiency of estimation.
- Permits greater balancing of statistical power of tests of differences between strata by sampling equal numbers from strata varying widely in size.

Disadvantages

- Requires selection of relevant stratification variables which can be difficult.
- Is not useful when there are no homogeneous subgroups.
- Can be expensive to implement.

Poststratification

Stratification is sometimes introduced after the sampling phase in a process called "poststratification". This approach is typically implemented due to a lack of prior knowledge of an appropriate stratifying variable or when the experimenter lacks the necessary information to create a stratifying variable during the sampling phase. Although the method is susceptible to the pitfalls of post hoc approaches, it can provide several benefits in the right situation. Implementation usually follows a simple random sample. In addition to allowing for stratification on an ancillary variable, poststratification can be used to implement weighting, which can improve the precision of a sample's estimates.

Oversampling

Choice-based sampling is one of the stratified sampling strategies. In choice-based sampling, the data are stratified on the target and a sample is taken from each stratum so that the rare target class will be more represented in the sample. The model is then built on this biased sample. The effects of the input variables on the target are often estimated with more precision with the choice-based sample even when a smaller overall sample size is taken, compared to a random sample. The results usually must be adjusted to correct for the oversampling.

Probability-Proportional-to-Size Sampling

In some cases the sample designer has access to an "auxiliary variable" or "size measure", believed to be correlated to the variable of interest, for each element in the population. These data can be used to improve accuracy in sample design. One option is to use the auxiliary variable as a basis for stratification, as discussed above.

Another option is probability proportional to size ('PPS') sampling, in which the selection probability for each element is set to be proportional to its size measure, up to a maximum of 1. In a simple PPS design, these selection probabilities can then be used as the basis for Poisson sampling. However, this has the drawback of variable sample size, and different portions of the population may still be over- or under-represented due to chance variation in selections.

Systematic sampling theory can be used to create a probability proportionate to size sample. This is done by treating each count within the size variable as a single sampling unit. Samples are then identified by selecting at even intervals among these counts within the size variable. This method is sometimes called PPS-sequential or monetary unit sampling in the case of audits or forensic sampling.

Example: Suppose we have six schools with populations of 150, 180, 200, 220, 260, and 490 students respectively (total 1500 students), and we want to use student population as the basis for a PPS sample of size three. To do this, we could allocate the first school numbers 1 to 150, the second school 151 to 330 (= 150 + 180), the third school 331 to 530, and so on to the last school (1011 to 1500). We then generate a random start between 1 and 500 (equal to 1500/3) and count through the school populations by multiples of 500. If our random start was 137, we would select the schools which have been allocated numbers 137, 637, and 1137, i.e. the first, fourth, and sixth schools.

The PPS approach can improve accuracy for a given sample size by concentrating sample on large elements that have the greatest impact on population estimates. PPS sampling is commonly used for surveys of businesses, where element size varies greatly and auxiliary information is often available—for instance, a survey attempting to measure the number of guest-nights spent in hotels might use each hotel's number of rooms as an auxiliary variable. In some cases, an older measurement of the variable of interest can be used as an auxiliary variable when attempting to produce more current estimates.

Cluster Sampling

Sometimes it is more cost-effective to select respondents in groups ('clusters'). Sampling is often clustered by geography, or by time periods. (Nearly all samples are in some sense 'clustered' in time – although this is rarely taken into account in the analysis.) For instance, if surveying households within a city, we might choose to select 100 city blocks and then interview every household within the selected blocks.

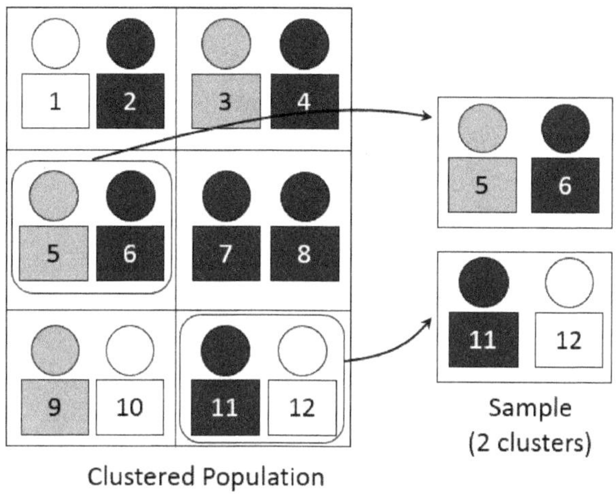

A visual representation of selecting a random sample using the cluster sampling technique

Clustering can reduce travel and administrative costs. In the example above, an interviewer can make a single trip to visit several households in one block, rather than having to drive to a different block for each household.

It also means that one does not need a sampling frame listing all elements in the target population. Instead, clusters can be chosen from a cluster-level frame, with an element-level frame created only for the selected clusters. In the example above, the sample only requires a block-level city map for initial selections, and then a household-level map of the 100 selected blocks, rather than a household-level map of the whole city.

Cluster sampling (also known as clustered sampling) generally increases the variability of sample estimates above that of simple random sampling, depending on how the clusters differ between one another as compared to the within-cluster variation. For this reason, cluster sampling requires a larger sample than SRS to achieve the same level of accuracy – but cost savings from clustering might still make this a cheaper option.

Cluster sampling is commonly implemented as multistage sampling. This is a complex form of cluster sampling in which two or more levels of units are embedded one in the other. The first stage consists of constructing the clusters that will be used to sample from. In the second stage, a sample of primary units is randomly selected from each cluster (rather than using all units contained in all selected clusters). In following stages, in each of those selected clusters, additional samples of units are selected, and so on. All ultimate units (individuals, for instance) selected at the last step of this procedure are then surveyed. This technique, thus, is essentially the process of taking random subsamples of preceding random samples.

Multistage sampling can substantially reduce sampling costs, where the complete population list would need to be constructed (before other sampling methods could be applied). By eliminating the work involved in describing clusters that are not selected, multistage sampling can reduce the large costs associated with traditional cluster sampling. However, each sample may not be a full representative of the whole population.

Quota Sampling

In quota sampling, the population is first segmented into mutually exclusive sub-groups, just as in stratified sampling. Then judgment is used to select the subjects or units from each segment based on a specified proportion. For example, an interviewer may be told to sample 200 females and 300 males between the age of 45 and 60.

It is this second step which makes the technique one of non-probability sampling. In quota sampling the selection of the sample is non-random. For example, interviewers might be tempted to interview those who look most helpful. The problem is that these samples may be biased because not everyone gets a chance of selection. This random element is its greatest weakness and quota versus probability has been a matter of controversy for several years.

Minimax Sampling

In imbalanced datasets, where the sampling ratio does not follow the population statistics, one can resample the dataset in a conservative manner called minimax sampling. The minimax sampling has its origin in Anderson minimax ratio whose value is proved to be 0.5: in a binary classification, the class-sample sizes should be chosen equally. This ratio can be proved to be minimax ratio only under the assumption of LDA classifier with Gaussian distributions. The notion of minimax sampling is recently developed for a general class of classification rules, called class-wise smart classifiers. In this case, the sampling ratio of classes is selected so that the worst case classifier error over all the possible population statistics for class prior probabilities, would be the

Accidental Sampling

Accidental sampling (sometimes known as grab, convenience or opportunity sampling) is a type of nonprobability sampling which involves the sample being drawn from that part of the population which is close to hand. That is, a population is selected because it is readily available and convenient. It may be through meeting the person or including a person in the sample when one meets them or chosen by finding them through technological means such as the internet or through phone. The researcher using such a sample cannot scientifically make generalizations about the total population from this sample because it would not be representative enough. For example, if the interviewer were to conduct such a survey at a shopping center early in the morning on a given day, the people that he/she could interview would be limited to those given there at that given time, which would not represent the views of other members of society in such an area, if the survey were to be conducted at different times of day and several times per week. This type of sampling is most useful for pilot testing. Several important considerations for researchers using convenience samples include:

- Are there controls within the research design or experiment which can serve to lessen the impact of a non-random convenience sample, thereby ensuring the results will be more representative of the population?
- Is there good reason to believe that a particular convenience sample would or should respond or behave differently than a random sample from the same population?
- Is the question being asked by the research one that can adequately be answered using a convenience sample?

In social science research, snowball sampling is a similar technique, where existing study subjects are used to recruit more subjects into the sample. Some variants of snowball sampling, such as respondent driven sampling, allow calculation of selection probabilities and are probability sampling methods under certain conditions.

Line-Intercept Sampling

Line-intercept sampling is a method of sampling elements in a region whereby an element is sampled if a chosen line segment, called a "transect", intersects the element.

Panel Sampling

Panel sampling is the method of first selecting a group of participants through a random sampling method and then asking that group for (potentially the same) information several times over a period of time. Therefore, each participant is interviewed at two or more time points; each period of data collection is called a "wave". The method was developed by sociologist Paul Lazarsfeld in 1938 as a means of studying political campaigns. This longitudinal sampling-method allows estimates of changes in the population, for example with regard to chronic illness to job stress to weekly food expenditures. Panel sampling can also be used to inform researchers about within-person health changes due to age or to help explain changes in continuous dependent variables such as spousal interaction. There have been several proposed methods of analyzing panel data, including MANOVA, growth curves, and structural equation modeling with lagged effects.

Snowball Sampling

Snowball sampling involves finding a small group of initial respondents and using them to recruit more respondents. It is particularly useful in cases where the population is hidden or difficult to enumerate.

Theoretical Sampling

Theoretical sampling occurs when samples are selected on the basis of the results of the data collected so far with a goal of developing a deeper understanding of the area or develop theory.

Replacement of Selected Units

Sampling schemes may be without replacement ('WOR'—no element can be selected more than once in the same sample) or with replacement ('WR'—an element may appear multiple times in the one sample). For example, if we catch fish, measure them, and immediately return them to the water before continuing with the sample, this is a WR design, because we might end up catching and measuring the same fish more than once. However, if we do not return the fish to the water (e.g., if we eat the fish), this becomes a WOR design.

Sample Size

Formulas, tables, and power function charts are well known approaches to determine sample size.

Steps for using sample size tables

- Postulate the effect size of interest, α, and β.
- Check sample size table
    1. Select the table corresponding to the selected α
    2. Locate the row corresponding to the desired power
    3. Locate the column corresponding to the estimated effect size.
    4. The intersection of the column and row is the minimum sample size required.

Sampling and Data Collection

Good data collection involves:

- Following the defined sampling process
- Keeping the data in time order
- Noting comments and other contextual events
- Recording non-responses

Applications of Sampling

Sampling enables the selection of right data points from within the larger data set to estimate the characteristics of the whole population. For example, there are about 600 million tweets produced every day. Is it necessary to look at all of them to determine the topics that are discussed during the day? Is it necessary to look at all the tweets to determine the sentiment on each of the topics? In manufacturing different types of sensory data such as acoustics, vibration, pressure, current, voltage and controller data are available at short time intervals. To predict down-time it may not be necessary to look at all the data but a sample may be sufficient.

A theoretical formulation for sampling Twitter data has been developed.

Errors in Sample Surveys

Survey results are typically subject to some error. Total errors can be classified into sampling errors and non-sampling errors. The term "error" here includes systematic biases as well as random errors.

Sampling Errors and Biases

Sampling errors and biases are induced by the sample design. They include:

- Selection bias: When the true selection probabilities differ from those assumed in calculating the results.
- Random sampling error: Random variation in the results due to the elements in the sample being selected at random.

Non-Sampling Error

Non-sampling errors are other errors which can impact the final survey estimates, caused by problems in data collection, processing, or sample design. They include:

- Over-coverage: Inclusion of data from outside of the population.
- Under-coverage: Sampling frame does not include elements in the population.
- Measurement error: e.g. when respondents misunderstand a question, or find it difficult to answer.
- Processing error: Mistakes in data coding.
- Non-response: Failure to obtain complete data from all selected individuals.

After sampling, a review should be held of the exact process followed in sampling, rather than that intended, in order to study any effects that any divergences might have on subsequent analysis. A particular problem is that of non-response.

Two major types of non-response exist: unit nonresponse (referring to lack of completion of any part of the survey) and item non-response (submission or participation in survey but failing to complete one or more components/questions of the survey). In survey sampling, many of the individuals identified as part of the sample may be unwilling to participate, not have the time to participate (opportunity cost), or survey administrators may not have been able to contact them. In this case, there is a risk of differences, between respondents and nonrespondents, leading to biased estimates of population parameters. This is often addressed by improving survey design, offering incentives, and conducting follow-up studies which make a repeated attempt to contact the unresponsive and to characterize their similarities and differences with the rest of the frame. The effects can also be mitigated by weighting the data when population benchmarks are available or by imputing data based on answers to other questions.

Nonresponse is particularly a problem in internet sampling. Reasons for this problem include improperly designed surveys, over-surveying (or survey fatigue), and the fact that potential participants hold multiple e-mail addresses, which they don't use anymore or don't check regularly.

Survey Weights

In many situations the sample fraction may be varied by stratum and data will have to be weighted to correctly represent the population. Thus for example, a simple random sample of individuals in the United Kingdom might include some in remote Scottish islands who would be inordinately expensive to sample. A cheaper method would be to use a stratified sample with urban and rural strata. The rural sample could be under-represented in the sample, but weighted up appropriately in the analysis to compensate.

More generally, data should usually be weighted if the sample design does not give each individual an equal chance of being selected. For instance, when households have equal selection probabilities but one person is interviewed from within each household, this gives people from large households a smaller chance of being interviewed. This can be accounted for using survey weights. Similarly, households with more than one telephone line have a greater chance of being selected in a random digit dialing sample, and weights can adjust for this.

Weights can also serve other purposes, such as helping to correct for non-response.

Methods of Producing Random Samples

- Random number table
- Mathematical algorithms for pseudo-random number generators
- Physical randomization devices such as coins, playing cards or sophisticated devices such as ERNIE

Standard Error

The standard error (SE) is the standard deviation of the sampling distribution of a statistic, most commonly of the mean.

The standard error of the mean (SEM) can be seen to depict the relationship between the dispersion of individual observations around the population mean (the standard deviation), and the dispersion of sample means around the population mean (the standard error). Different samples drawn from that same population would in general have different values of the sample mean, so there is a distribution of sampled means (with its own mean and variance). The relationship with the standard deviation is defined such that, for a given sample size, the standard error equals the standard deviation divided by the square root of the sample size. As the sample size increases, the dispersion of the sample means clusters more closely around the population mean and the standard error decreases.

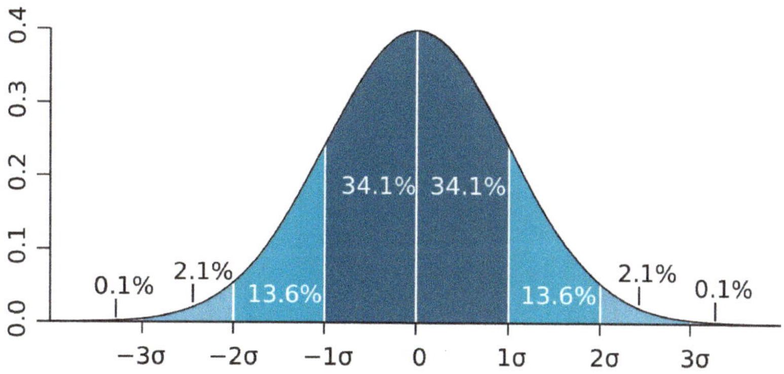

For a value that is sampled with an unbiased normally distributed error, the above depicts the proportion of samples that would fall between 0, 1, 2, and 3 standard deviations above and below the actual value.

In regression analysis, the term "standard error" is also used in the phrase standard error of the regression to mean the ordinary least squares estimate of the standard deviation of the underlying errors. The term may also be used to refer to an estimate of that standard deviation, derived from a particular sample used to compute the estimate.

Introduction to the Standard Error

The standard error is a quantitative measure of uncertainty. Consider the following scenarios.

Scenario 1. For an upcoming national election, 2000 voters are chosen at random and asked if they will vote for candidate A or candidate B. Of the 2000 voters, 1040 (52%) state that they will vote for candidate A. The researchers report that candidate A is expected to receive 52% of the final vote, with a margin of error of 2%. In this scenario, the 2000 voters are a sample from all the actual voters. The sample proportion of 52% is an estimate of the true proportion who will vote for candidate A in the actual election. The margin of error of 2% is a quantitative measure of the uncertainty – the possible difference between the true proportion who will vote for candidate A and the estimate of 52%.

Scenario 2. A medical research team tests a new drug to lower cholesterol. They report that, in a sample of 400 patients, the new drug lowers cholesterol by an average of 20 units (mg/dL). The 95% confidence interval for the average effect of the drug is that it lowers cholesterol by 18 to 22 units. In this scenario, the 400 patients are a sample of all patients who may be treated with the drug. The confidence interval of 18 to 22 is a quantitative measure of the uncertainty – the possible difference between the true average effect of the drug and the estimate of 20 mg/dL.

In each of these scenarios, a sample of observations is drawn from a large population. The proportion or the mean is calculated using the sample. Because of random variation in sampling, the proportion or mean calculated using the sample will usually differ from the true proportion or mean in the entire population. A quantitative measure of uncertainty is reported: a margin of error of 2%, or a confidence interval of 18 to 22. The margin of error and the confidence interval are based on a quantitative measure of uncertainty: the standard error. The standard error of a proportion and the standard error of the mean describe the possible variability of the estimated value based on the sample around the true proportion or true mean

Standard errors are used in many hypothesis tests, such as t-tests. They may be used to calculate confidence intervals.

Standard Error of the Mean (SEM)

This section will focus on the standard error of the mean. Later sections will present the standard error of other statistics, such as the standard error of a proportion, the standard error of the difference of two means, the standard error of the difference of two proportions and so on.

The concept of a sampling distribution is key to understanding the standard error. Two data sets will be helpful to illustrate the concept of a sampling distribution and its use to calculate the standard error. As will be shown, the standard error is the standard deviation of the sampling distribution.

Sampling from a Distribution with a Large Standard Deviation

The first data set consists of the ages of 9,732 women who completed the 2012 Cherry Blossom run, a 10-mile race held in Washington each spring. The age data are in the data set run10 from the R package openintro that accompanies the textbook by Dietz

Larger Sample Sizes Give Smaller Standard Errors

As would be expected, larger sample sizes give smaller standard errors. The graphs below show the sampling distribution of the mean for samples of size 4, 9, and 25. As the sample size increases, the sampling distribution become more narrow, and the standard error decreases.

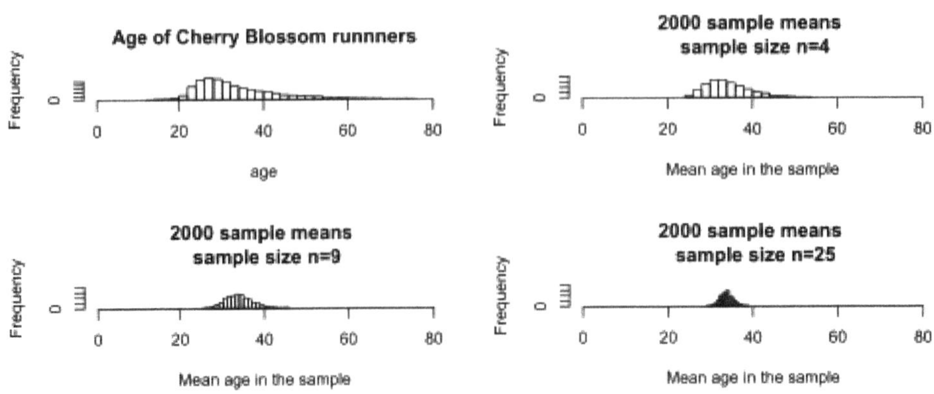

Using a Sample to Estimate the Standard Error

In the examples so far, the population standard deviation σ was assumed to be known. If σ is known, the standard error is calculated using the formula

$$\sigma_{\bar{x}} = \frac{\sigma}{\sqrt{n}}$$

where

σ is the standard deviation of the population.

n is the size (number of observations) of the sample.

It is rare that the true population standard deviation is known. However, the sample standard deviation, s, is an estimate of σ. If σ is not known, the standard error is estimated using the formula

$$s_{\bar{x}} = \frac{s}{\sqrt{n}}$$

where

s is the sample standard deviation

n is the size (number of observations) of the sample.

Notice that $s_{\bar{x}} = \frac{s}{\sqrt{n}}$ is only an estimate of the true standard error, $\sigma_{\bar{x}} = \frac{\sigma}{\sqrt{n}}$.

In an example above, n=16 runners were selected at random from the 9,732 runners. The ages in that sample were 23, 27, 28, 29, 31, 31, 32, 33, 34, 38, 40, 40, 48, 53, 54, and 55. The standard deviation of the age for the 16 runners is 10.23, which is somewhat greater than the true population standard deviation σ = 9.27 years. Compare the true standard error of the mean to the standard error estimated using this sample.

The true standard error of the mean, using σ = 9.27, is

$$\sigma_{\bar{x}} = \frac{\sigma}{\sqrt{n}} = \frac{9.27}{\sqrt{16}} = 2.32$$

The standard error of the mean estimated by using the sample standard deviation, s = 10.23, is

$$s_{\bar{x}} = \frac{s}{\sqrt{n}} = \frac{10.23}{\sqrt{16}} = 2.56$$

The true standard error of the mean is 2.32. The standard error estimated using the sample standard deviation is 2.56.

For the purpose of hypothesis testing or estimating confidence intervals, the standard error is primarily of use when the sampling distribution is normally distributed, or approximately normally distributed. These assumptions may be approximately met when the population from which samples are taken is normally distributed, or when the sample size is sufficiently large to rely on the Central Limit Theorem.

Standard Error of the Mean

The standard error of the mean (SEM) is the standard deviation of the sample-mean's estimate of a population mean. (It can also be viewed as the standard deviation of the error in the sample mean with respect to the true mean, since the sample mean is an unbiased estimator.) SEM is usually estimated by the sample estimate of the population standard deviation (sample standard deviation) divided by the square root of the sample size (assuming statistical independence of the values in the sample):

$$\mathrm{SE}_{\bar{x}} = \frac{s}{\sqrt{n}}$$

where

s is the sample standard deviation (i.e., the sample-based estimate of the standard deviation of the population), and

n is the size (number of observations) of the sample.

This estimate may be compared with the formula for the true standard deviation of the sample mean:

$$\mathrm{SD}_{\bar{x}} = \frac{\sigma}{\sqrt{n}}$$

where

σ is the standard deviation of the population.

This formula may be derived from what we know about the variance of a sum of independent random variables.

- If $X_1, X_2, \ldots, X_n$ are n independent observations from a population that has a mean μ and standard deviation σ, then the variance of the total $T = (X_1 + X_2 + \cdots + X_n)$ is $n\sigma^2$.
- The variance of T/n must be $\frac{1}{n^2} n\sigma^2 = \frac{\sigma^2}{n}$.
- And the standard deviation of T/n must be $\sigma/\sqrt{n}$.
- Of course, T/n is the sample mean $\bar{x}$.

Note: the standard error and the standard deviation of small samples tend to systematically underestimate the population standard error and deviations: the standard error of the mean is a biased estimator of the population standard error. With n = 2 the underestimate is about 25%, but for n = 6 the underestimate is only 5%. Gurland and Tripathi (1971) provide a correction and equation for this effect. Sokal and Rohlf (1981) give an equation of the correction factor for small samples of n < 20.

A practical result: Decreasing the uncertainty in a mean value estimate by a factor of two requires acquiring four times as many observations in the sample. Or decreasing standard error by a factor of ten requires a hundred times as many observations.

Student Approximation when σ Value is Unknown

In many practical applications, the true value of σ is unknown. As a result, we need to use a distribution that takes into account that spread of possible σ's. When the true underlying distribution is known to be Gaussian, although with unknown σ, then the resulting estimated distribution follows the Student t-distribution. The standard error is the standard deviation of the Student t-distribution. T-distributions are slightly different from Gaussian, and vary depending on the size of the sample. To estimate the standard error of a student t-distribution it is sufficient to use the sample standard deviation «s» instead of σ, and we could use this value to calculate confidence intervals.

Note: The Student's probability distribution is a good approximation of the Gaussian when the sample size is over 100.

Assumptions and Usage

If its sampling distribution is normally distributed, the sample mean, its standard error, and the quantiles of the normal distribution can be used to calculate confidence intervals for the mean. The following expressions can be used to calculate the upper and lower 95% confidence limits, where $\bar{x}$ is equal to the sample mean, SE is equal to the standard error for the sample mean, and 1.96 is the 0.975 quantile of the normal distribution:

$$\text{Upper 95\% limit} = \bar{x} + (\mathbf{SE} \times \mathbf{1.96}), \text{ and}$$
$$\text{Lower 95\% limit} = \bar{x} - (\mathbf{SE} \times \mathbf{1.96}).$$

In particular, the standard error of a sample statistic (such as sample mean) is the estimated standard deviation of the error in the process by which it was generated. In other words, it is the standard deviation of the sampling distribution of the sample statistic. The notation for standard error can be any one of SE, SEM (for standard error of measurement or mean), or S_E.

Standard errors provide simple measures of uncertainty in a value and are often used because:

- If the standard error of several individual quantities is known then the standard error of some function of the quantities can be easily calculated in many cases;
- Where the probability distribution of the value is known, it can be used to calculate a good approximation to an exact confidence interval; and
- Where the probability distribution is unknown, relationships like Chebyshev's or the Vysochanskiï–Petunin inequality can be used to calculate a conservative confidence interval
- As the sample size tends to infinity the central limit theorem guarantees that the sampling distribution of the mean is asymptotically normal.

Standard Error of Mean Versus Standard Deviation

In scientific and technical literature, experimental data are often summarized either using the mean and standard deviation or the mean with the standard error. This often leads to confusion about their interchangeability. However, the mean and standard deviation are descriptive statistics, whereas the standard error of the mean describes bounds on a random sampling process. Despite the small difference in equations for the standard deviation and the standard error, this small difference changes the meaning of what is being reported from a description of the variation in measurements to a probabilistic statement about how the number of samples will provide a better bound on estimates of the population mean, in light of the central limit theorem.

Put simply, the standard error of the sample mean is an estimate of how far the sample mean is likely to be from the population mean, whereas the standard deviation of the sample is the degree to which individuals within the sample differ from the sample mean. If the population standard deviation is finite, the standard error of the mean of the sample will tend to zero with increasing sample size, because the estimate of the population mean will improve, while the standard deviation of the sample will tend to approximate the population standard deviation as the sample size increases.

Correction for Finite Population

The formula given above for the standard error assumes that the sample size is much smaller than the population size, so that the population can be considered to be effectively infinite in size. This is usually the case even with finite populations, because most of the time, people are primarily interested in managing the processes that created the existing finite population; this is called an analytic study, following W. Edwards Deming. If people are interested in managing an existing finite population that will not change over time, then it is necessary to adjust for the population size; this is called an enumerative study.

When the sampling fraction is large (approximately at 5% or more) in an enumerative study, the estimate of the standard error must be corrected by multiplying by a "finite population correction"

$$\mathrm{FPC} = \sqrt{\frac{N-n}{N-1}}$$

to account for the added precision gained by sampling close to a larger percentage of the population. The effect of the FPC is that the error becomes zero when the sample size n is equal to the population size N.

Correction for Correlation in the Sample

If values of the measured quantity A are not statistically independent but have been obtained from known locations in parameter space **x**, an unbiased estimate of the true standard error of the mean (actually a correction on the standard deviation part) may be obtained by multiplying the calculated standard error of the sample by the factor f:

$$f = \sqrt{\frac{1+\rho}{1-\rho}},$$

where the sample bias coefficient ρ is the widely used Prais-Winsten estimate of the autocorrelation-coefficient (a quantity between −1 and +1) for all sample point pairs.

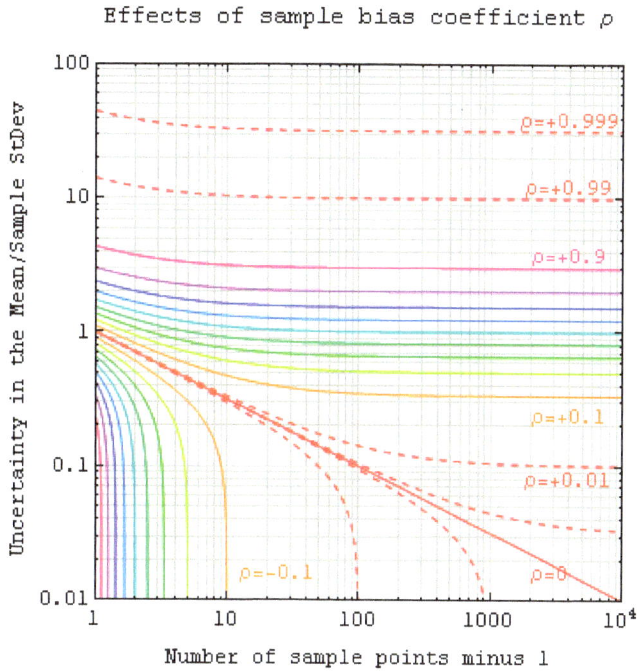

Expected error in the mean of A for a sample of n data points with sample bias coefficient ρ. The unbiased **standard error** plots as the ρ=0 diagonal line with log-log slope -½.

This approximate formula is for moderate to large sample sizes; the reference gives the exact formulas for any sample size, and can be applied to heavily autocorrelated time series like Wall Street stock quotes. Moreover, this formula works for positive and negative ρ alike.

Relative Standard Error

The relative standard error of a sample mean is the standard error divided by the mean and expressed as a percentage. It can only be calculated if the mean is a non-zero value.

As an example of the use of the relative standard error, consider two surveys of household income that both result in a sample mean of $50,000. If one survey has a standard error of $10,000 and the other has a standard error of $5,000, then the relative standard errors are 20% and 10% respectively. The survey with the lower relative standard error can be said to have a more precise measurement, since it has proportionately less sampling variation around the mean. In fact, data organizations often set reliability standards that their data must reach before publication. For example, the U.S. National Center for Health Statistics typically does not report an estimated mean if its relative standard error exceeds 30%. (NCHS also typically requires at least 30 observations – if not more – for an estimate to be reported.)

REFERENCES

1. Lescroël, A. L.; Ballard, G.; Grémillet, D.; Authier, M.; Ainley, D. G. (2014). Descamps, Sébastien, ed. "Antarctic Climate Change: Extreme Events Disrupt Plastic Phenotypic Response in Adélie Penguins". PLoS ONE. 9 (1): e85291. doi:10.1371/journal.pone.0085291. PMC 3906005Freely accessible. PMID 24489657.
2. Data Collection and Analysis By Dr. Roger Sapsford, Victor Jupp ISBN 0-7619-5046-X
3. Weimer, J. (ed.) (1995). Research Techniques in Human Engineering. Englewood Cliffs, NJ: Prentice Hall ISBN 0-13-097072-7
4. "Glossary of statistical terms: Population". Statistics.com. Retrieved 22 February 2016.
5. Weisstein, Eric W. "Statistical population". MathWorld.
6. Yates, Daniel S.; Moore, David S; Starnes, Daren S. (2003). The Practice of Statistics (2nd ed.). New York: Freeman. ISBN 978-0-7167-4773-4.
7. "Glossary of statistical terms: Sample". Statistics.com. Retrieved 22 February 2016.
8. Kelley, Ken; Preacher, Kristopher J. (2012). "On Effect Size". Psychological Methods. 17 (2): 137–152. doi:10.1037/a0028086.
9. Wilkinson, Leland (1999). "Statistical methods in psychology journals: Guidelines and explanations". American Psychologist. 54 (8): 594–604. doi:10.1037/0003-066X.54.8.594.
10. Nakagawa, Shinichi; Cuthill, Innes C (2007). "Effect size, confidence interval and statistical significance: a practical guide for biologists". Biological Reviews Cambridge Philosophical Society. 82 (4): 591–605. doi:10.1111/j.1469-185X.2007.00027.x. PMID 17944619.
11. Ellis, Paul D. (2010). The Essential Guide to Effect Sizes: Statistical Power, Meta-Analysis, and the Interpretation of Research Results. Cambridge University Press. ISBN 978-0-521-14246-5.
12. Brand A, Bradley MT, Best LA, Stoica G (2008). "Accuracy of effect size estimates from published psychological research" (PDF). Perceptual and Motor Skills. 106 (2): 645–649. doi:10.2466/PMS.106.2.645-649. PMID 18556917.
13. Brand A, Bradley MT, Best LA, Stoica G (2011). "Multiple trials may yield exaggerated effect size estimates" (PDF). The Journal of General Psychology. 138 (1): 1–11. doi:10.1080/00221309.2010.520360.
14. Cohen, Jacob (1988). Statistical Power Analysis for the Behavioral Sciences. Routledge. ISBN 1-134-74270-3.
15. Cohen, J (1992). "A power primer". Psychological Bulletin. 112 (1): 155–159. doi:10.1037/0033-2909.112.1.155. PMID 19565683.
16. Olejnik, S. & Algina, J. 2003. Generalized Eta and Omega Squared Statistics: Measures of Effect Size for Some Common Research Designs Psychological

Methods. 8:(4)434-447. http://cps.nova.edu/marker/olejnik2003.pdf

17. Steiger, J. H. 2004. Beyond the F test: Effect size confidence intervals and tests of close fit in the analysis of variance and contrast analysis. Psychological Methods 9:(2) 164-182. http://www.statpower.net/Steiger%20Biblio/Steiger04.pdf
18. Hair/Hult/Ringle/Sarstedt, A Primer on Partial Least Squares Structural Equation Modeling (PLS-SEM), Sage, 2014, pp. 177-178.
19. Larry V. Hedges & Ingram Olkin (1985). Statistical Methods for Meta-Analysis. Orlando: Academic Press. ISBN 0-12-336380-2.
20. Sawilowsky, S (2009). “New effect size rules of thumb.”. Journal of Modern Applied Statistical Methods. 8 (2): 467–474.
21. Robert E. McGrath; Gregory J. Meyer (2006). “When Effect Sizes Disagree: The Case of r and d” (PDF). Psychological Methods. 11 (4): 386–401. doi:10.1037/1082-989x.11.4.386.
22. Hartung, Joachim; Knapp, Guido; Sinha, Bimal K. (2008). Statistical Meta-Analysis with Applications. John Wiley & Sons. ISBN 978-1-118-21096-3.
23. Kenny, David A. (1987). “Chapter 13” (PDF). Statistics for the Social and Behavioral Sciences. Little, Brown. ISBN 978-0-316-48915-7.
24. Larry V. Hedges (1981). “Distribution theory for Glass’ estimator of effect size and related estimators”. Journal of Educational Statistics. 6 (2): 107–128. doi:10.3102/10769986006002107.
25. Aaron, B., Kromrey, J. D., & Ferron, J. M. (1998, November). Equating r-based and d-based effect-size indices: Problems with a commonly recommended formula. Paper presented at the annual meeting of the Florida Educational Research Association, Orlando, FL. (ERIC Document Reproduction Service No. ED433353)
26. Sheskin, David J. (2003). Handbook of Parametric and Nonparametric Statistical Procedures (Third ed.). CRC Press. ISBN 978-1-4200-3626-8.
27. Deeks J (1998). “When can odds ratios mislead? : Odds ratios should be used only in case-control studies and logistic regression analyses”. BMJ. 317 (7166): 1155–6. doi:10.1136/bmj.317.7166.1155a. PMC 1114127Freely accessible. PMID 9784470.
28. Stegenga, J. (2015). “Measuring Effectiveness”. Studies in History and Philosophy of Biological and Biomedical Sciences. 54: 62–71. doi:10.1016/j.shpsc.2015.06.003.
29. McGraw KO, Wong SP (1992). “A common language effect size statistic”. Psychological Bulletin. 111 (2): 361–365. doi:10.1037/0033-2909.111.2.361.
30. McGraw KO, Wong SP (1992). “A common language effect size statistic”. Psychological Bulletin. 111 (2): 381. doi:10.1037/0033-2909.111.2.361.
31. Kerby, D. S. (2014). The simple difference formula: An approach to teaching nonparametric correlation. Comprehensive Psychology, volume 3, article 1.

doi:10.2466/11.IT.3.1. link to article

32. Grissom RJ (1994). "Statistical analysis of ordinal categorical status after therapies". Journal of Consulting and Clinical Psychology. 62 (2): 281–284. doi:10.1037/0022-006X.62.2.281.
33. Vargha, András; Delaney, Harold D. (2000). "A Critique and Improvement of the CL Common Language Effect Size Statistics of McGraw and Wong" (PDF). Journal of Educational and Behavioral Statistics. 25 (2): 101–132. doi:10.3102/10769986025002101.
34. Cureton, E.E. (1956). "Rank-biserial correlation". Psychometrika. 21 (3): 287–290. doi:10.1007/BF02289138.
35. Wendt, H. W. (1972). "Dealing with a common problem in social science: A simplified rank-biserial coefficient of correlation based on the U statistic". European Journal of Social Psychology. 2 (4): 463–465. doi:10.1002/ejsp.2420020412.
36. Cliff, Norman (1993). "Dominance statistics: Ordinal analyses to answer ordinal questions". Psychological Bulletin. 114 (3): 494. doi:10.1037/0033-2909.114.3.494.
37. Russell V. Lenth. "Java applets for power and sample size". Division of Mathematical Sciences, the College of Liberal Arts or The University of Iowa. Retrieved 2008-10-08.
38. Lipsey, M.W.; et al. (2012). Translating the Statistical Representation of the Effects of Education Interventions Into More Readily Interpretable Forms (PDF). United States: U.S. Dept of Education, National Center for Special Education Research, Institute of Education Sciences,NCSER 2013-3000.
39. Messner SF (1992). "Exploring the Consequences of Erratic Data Reporting for Cross-National Research on Homicide". Journal of Quantitative Criminology. 8 (2): 155–173. doi:10.1007/bf01066742.
40. Hand, David J.; Adèr, Herman J.; Mellenbergh, Gideon J. (2008). Advising on Research Methods: A Consultant's Companion. Huizen, Netherlands: Johannes van Kessel. pp. 305–332. ISBN 90-79418-01-3.
41. Mohan, Karthika; Pearl, Judea; Tian, Jin (2013). Advances in Neural Information Processing Systems 26. pp. 1277–1285.
42. Karvanen, Juha (2015). "Study design in causal models". Scandinavian Journal of Statistics. 42 (2): 361–377. doi:10.1111/sjos.12110.
43. Polit DF Beck CT (2012). Nursing Research: Generating and Assessing Evidence for Nursing Practice, 9th ed. Philadelphia, USA: Wolters Klower Health, Lippincott Williams & Wilkins.
44. Deng. "On Biostatistics and Clinical Trials". Retrieved 13 May 2016.
45. http://missingdata.lshtm.ac.uk/index.php?option=com_content&view=article&id=76%3Amissing-at-random-mar&catid=40%3Amissingness-mechanisms&Itemid=96
46. Little, Roderick (2002). Statistical analysis with missing data. Hoboken, N.J: Wiley. ISBN 978-0471183860.

47. Stoop, I.; Billiet, J.; Koch, A.; Fitzgerald, R. (2010). Reducing Survey Nonresponse: Lessons Learned from the European Social Survey. Oxford: Wiley-Blackwell. ISBN 0-470-51669-0.
48. Graham J.W.; Olchowski A.E.; Gilreath T.D. (2007). "How Many Imputations Are Really Needed? Some Practical Clarifications of Multiple Imputation Theory". Preventative Science. 8 (3): 208–213. doi:10.1007/s11121-007-0070-9.
49. Mohan, Karthika; Pearl, Judea (2014). "On the testability of models with missing data". Proceedings of AISTAT-2014, Forthcoming.
50. Darwiche, Adnan (2009). Modeling and Reasoning with Bayesian Networks. Cambridge University Press.
51. Potthoff, R.F.; Tudor, G.E.; Pieper, K.S.; Hasselblad, V. (2006). "Can one assess whether missing data are missing at random in medical studies?". Statistical Methods in Medical Research. 15 (3): 213–234. doi:10.1191/0962280206sm448oa.
52. Pearl, Judea; Mohan, Karthika (2013). Recoverability and Testability of Missing data: Introduction and Summary of Results (PDF) (Technical report). UCLA Computer Science Department, R-417.
53. Mohan, K.; Van den Broeck, G.; Choi, A.; Pearl, J. (2014). "An Efficient Method for Bayesian Network Parameter Learning from Incomplete Data". Presented at Causal Modeling and Machine learning Workshop, ICML-2014.
54. Mirkes, E.M.; Coats, T.J.; Levesley, J.; Gorban, A.N. (2016). "Handling missing data in large healthcare dataset: A case study of unknown trauma outcomes". Computers in Biology and Medicine. 75: 203–216. doi:10.1016/j.compbiomed.2016.06.004.

CHAPTER 4

Statistical Inference

Chapter Contents

- **Introduction**
- **Statistical Assumption**
- **Estimation Theory**
- **Statistical Hypothesis Testing**

Statistical inference is the process of deducing properties of an underlying distribution by analysis of data. Inferential statistical analysis infers properties about a population: this includes testing hypotheses and deriving estimates. The population is assumed to be larger than the observed data set; in other words, the observed data is assumed to be sampled from a larger population.

Inferential statistics can be contrasted with descriptive statistics. Descriptive statistics is solely concerned with properties of the observed data, and does not assume that the data came from a larger population.

INTRODUCTION

Statistical inference makes propositions about a population, using data drawn from the population with some form of sampling. Given a hypothesis about a population, for which we wish to draw inferences, statistical inference consists of (firstly) selecting a statistical model of the process that generates the data and (secondly) deducing propositions from the model. Konishi & Kitagawa state, "The majority of the problems in statistical inference can be considered to be problems related to statistical modeling". Relatedly, Sir David

Cox has said, "How [the] translation from subject-matter problem to statistical model is done is often the most critical part of an analysis".

The conclusion of a statistical inference is a statistical proposition. Some common forms of statistical proposition are the following:

- a point estimate, i.e. a particular value that best approximates some parameter of interest;
- an interval estimate, e.g. a confidence interval (or set estimate), i.e. an interval constructed using a dataset drawn from a population so that, under repeated sampling of such datasets, such intervals would contain the true parameter value with the probability at the stated confidence level;
- a credible interval, i.e. a set of values containing, for example, 95% of posterior belief;
- rejection of a hypothesis;
- clustering or classification of data points into groups.

Models and Assumptions

Any statistical inference requires some assumptions. A statistical model is a set of assumptions concerning the generation of the observed data and similar data. Descriptions of statistical models usually emphasize the role of population quantities of interest, about which we wish to draw inference. Descriptive statistics are typically used as a preliminary step before more formal inferences are drawn.

Degree of Models/Assumptions

Statisticians distinguish between three levels of modeling assumptions;

- Fully parametric: The probability distributions describing the data-generation process are assumed to be fully described by a family of probability distributions involving only a finite number of unknown parameters. For example, one may assume that the distribution of population values is truly Normal, with unknown mean and variance, and that datasets are generated by 'simple' random sampling. The family of generalized linear models is a widely used and flexible class of parametric models.
- Non-parametric: The assumptions made about the process generating the data are much less than in parametric statistics and may be minimal. For example, every continuous probability distribution has a median, which may be estimated using the sample median or the Hodges–Lehmann–Sen estimator, which has good properties when the data arise from simple random sampling.
- Semi-parametric: This term typically implies assumptions 'in between' fully and non-parametric approaches. For example, one may assume that a population distribution has a finite mean. Furthermore, one may assume that the mean response level in the population depends in a truly linear manner on some covariate (a parametric assumption) but not make any parametric assumption

describing the variance around that mean (i.e. about the presence or possible form of any heteroscedasticity). More generally, semi-parametric models can often be separated into 'structural' and 'random variation' components. One component is treated parametrically and the other non-parametrically. The well-known Cox model is a set of semi-parametric assumptions.

Importance of Valid Models/Assumptions

Whatever level of assumption is made, correctly calibrated inference in general requires these assumptions to be correct; i.e. that the data-generating mechanisms really have been correctly specified.

Incorrect assumptions of 'simple' random sampling can invalidate statistical inference. More complex semi- and fully parametric assumptions are also cause for concern. For example, incorrectly assuming the Cox model can in some cases lead to faulty conclusions. Incorrect assumptions of Normality in the population also invalidates some forms of regression-based inference. The use of **any** parametric model is viewed skeptically by most experts in sampling human populations: "most sampling statisticians, when they deal with confidence intervals at all, limit themselves to statements about [estimators] based on very large samples, where the central limit theorem ensures that these [estimators] will have distributions that are nearly normal." In particular, a normal distribution "would be a totally unrealistic and catastrophically unwise assumption to make if we were dealing with any kind of economic population." Here, the central limit theorem states that the distribution of the sample mean "for very large samples" is approximately normally distributed, if the distribution is not heavy tailed.

Approximate Distributions

Given the difficulty in specifying exact distributions of sample statistics, many methods have been developed for approximating these.

With finite samples, approximation results measure how close a limiting distribution approaches the statistic's sample distribution: For example, with 10,000 independent samples the normal distribution approximates (to two digits of accuracy) the distribution of the sample mean for many population distributions, by the Berry–Esseen theorem. Yet for many practical purposes, the normal approximation provides a good approximation to the sample-mean's distribution when there are 10 (or more) independent samples, according to simulation studies and statisticians' experience. Following Kolmogorov's work in the 1950s, advanced statistics uses approximation theory and functional analysis to quantify the error of approximation. In this approach, the metric geometry of probability distributions is studied; this approach quantifies approximation error with, for example, the Kullback–Leibler divergence, Bregman divergence, and the Hellinger distance.

With indefinitely large samples, limiting results like the central limit theorem describe the sample statistic's limiting distribution, if one exists. Limiting results are not statements about finite samples, and indeed are irrelevant to finite samples. However, the asymptotic theory of limiting distributions is often invoked for work with finite samples. For example, limiting results are often invoked to justify the generalized method of moments and the use of generalized estimating equations, which are popular in econometrics and biostatistics. The magnitude of the difference between the limiting distribution and the true distribution (formally, the 'error' of the approximation) can be assessed using simulation. The heuristic application of limiting results to finite samples is common practice in many applications, especially with low-dimensional models with log-concave likelihoods (such as with one-parameter exponential families).

Randomization-Based Models

For a given dataset that was produced by a randomization design, the randomization distribution of a statistic (under the null-hypothesis) is defined by evaluating the test statistic for all of the plans that could have been generated by the randomization design. In frequentist inference, randomization allows inferences to be based on the randomization distribution rather than a subjective model, and this is important especially in survey sampling and design of experiments. Statistical inference from randomized studies is also more straightforward than many other situations. In Bayesian inference, randomization is also of importance: in survey sampling, use of sampling without replacement ensures the exchangeability of the sample with the population; in randomized experiments, randomization warrants a missing at random assumption for covariate information.

Objective randomization allows properly inductive procedures. Many statisticians prefer randomization-based analysis of data that was generated by well-defined randomization procedures. (However, it is true that in fields of science with developed theoretical knowledge and experimental control, randomized experiments may increase the costs of experimentation without improving the quality of inferences.) Similarly, results from randomized experiments are recommended by leading statistical authorities as allowing inferences with greater reliability than do observational studies of the same phenomena. However, a good observational study may be better than a bad randomized experiment.

The statistical analysis of a randomized experiment may be based on the randomization scheme stated in the experimental protocol and does not need a subjective model.

However, at any time, some hypotheses cannot be tested using objective statistical models, which accurately describe randomized experiments or random samples. In some cases, such randomized studies are uneconomical or unethical.

Model-Based Analysis of Randomized Experiments

It is standard practice to refer to a statistical model, often a linear model, when analyzing data from randomized experiments. However, the randomization scheme guides the choice of a statistical model. It is not possible to choose an appropriate model without knowing the randomization scheme. Seriously misleading results can be obtained analyzing data from randomized experiments while ignoring the experimental protocol; common mistakes include forgetting the blocking used in an experiment and confusing repeated measurements on the same experimental unit with independent replicates of the treatment applied to different experimental units.

Paradigms for Inference

Different schools of statistical inference have become established. These schools—or "paradigms"—are not mutually exclusive, and methods that work well under one paradigm often have attractive interpretations under other paradigms.

Bandyopadhyay & Forster describe four paradigms: "(i) classical statistics or error statistics, (ii) Bayesian statistics, (iii) likelihood-based statistics, and (iv) the Akaikean-Information Criterion-based statistics". The classical (or frequentist) paradigm, the Bayesian paradigm, and the AIC-based paradigm are summarized below. The likelihood-based paradigm is essentially a sub-paradigm of the AIC-based paradigm.

Frequentist Inference

This paradigm calibrates the plausibility of propositions by considering (notional) repeated sampling of a population distribution to produce datasets similar to the one at hand. By considering the dataset's characteristics under repeated sampling, the frequentist properties of a statistical proposition can be quantified—although in practice this quantification may be challenging.

Examples of frequentist inference

- p-value
- Confidence interval

Frequentist Inference, Objectivity, and Decision Theory

One interpretation of frequentist inference (or classical inference) is that it is applicable only in terms of frequency probability; that is, in terms of repeated sampling from a population. However, the approach of Neyman develops these procedures in terms of pre-experiment probabilities. That is, before undertaking an experiment, one decides on a rule for coming to a conclusion such that the probability of being correct is controlled in a suitable way: such a probability need not have a frequentist or repeated sampling interpretation. In contrast, Bayesian inference works in terms of conditional probabilities (i.e. probabilities conditional on the observed data), compared to the marginal (but conditioned on unknown parameters) probabilities used in the frequentist approach.

The frequentist procedures of significance testing and confidence intervals can be constructed without regard to utility functions. However, some elements of frequentist statistics, such as statistical decision theory, do incorporate utility functions. In particular, frequentist developments of optimal inference (such as minimum-variance unbiased estimators, or uniformly most powerful testing) make use of loss functions, which play the role of (negative) utility functions. Loss functions need not be explicitly stated for statistical theorists to prove that a statistical procedure has an optimality property. However, loss-functions are often useful for stating optimality properties: for example, median-unbiased estimators are optimal under absolute value loss functions, in that they minimize expected loss, and least squares estimators are optimal under squared error loss functions, in that they minimize expected loss.

While statisticians using frequentist inference must choose for themselves the parameters of interest, and the estimators/test statistic to be used, the absence of obviously explicit utilities and prior distributions has helped frequentist procedures to become widely viewed as 'objective'.

Bayesian Inference

The Bayesian calculus describes degrees of belief using the 'language' of probability; beliefs are positive, integrate to one, and obey probability axioms. Bayesian inference uses the available posterior beliefs as the basis for making statistical propositions. There are several different justifications for using the Bayesian approach.

Examples of Bayesian inference

- Credible interval for interval estimation
- Bayes factors for model comparison

Bayesian Inference, Subjectivity and Decision Theory

Many informal Bayesian inferences are based on "intuitively reasonable" summaries of the posterior. For example, the posterior mean, median and mode, highest posterior density intervals, and Bayes Factors can all be motivated in this way. While a user's utility function need not be stated for this sort of inference, these summaries do all depend (to some extent) on stated prior beliefs, and are generally viewed as subjective conclusions. (Methods of prior construction which do not require external input have been proposed but not yet fully developed.)

Formally, Bayesian inference is calibrated with reference to an explicitly stated utility, or loss function; the 'Bayes rule' is the one which maximizes expected utility, averaged over the posterior uncertainty. Formal Bayesian inference therefore automatically provides optimal decisions in a decision theoretic sense. Given assumptions, data and utility, Bayesian inference can be made for essentially any problem, although not every statistical inference need have a Bayesian interpretation. Analyses which are not formally Bayesian can be (logically) incoherent; a feature of Bayesian procedures which use proper priors (i.e. those integrable to one) is that they are guaranteed to be coherent. Some advocates of Bayesian inference assert that inference must take place in this decision-theoretic framework, and that Bayesian inference should not conclude with the evaluation and summarization of posterior beliefs.

AIC-Based Inference

Minimum Description Length

The minimum description length (MDL) principle has been developed from ideas in information theory and the theory of Kolmogorov complexity. The (MDL) principle selects statistical models that maximally compress the data; inference proceeds without assuming counterfactual or non-falsifiable "data-generating mechanisms" or probability models for the data, as might be done in frequentist or Bayesian approaches.

However, if a "data generating mechanism" does exist in reality, then according to Shannon's source coding theorem it provides the MDL description of the data, on average and asymptotically. In minimizing description length (or descriptive complexity), MDL estimation is similar to maximum likelihood estimation and maximum a posteriori estimation (using maximum-entropy Bayesian priors). However, MDL avoids assuming that the underlying probability model is known; the MDL principle can also be applied without assumptions that e.g. the data arose from independent sampling.

The MDL principle has been applied in communication-coding theory in information theory, in linear regression, and in data mining.

The evaluation of MDL-based inferential procedures often uses techniques or criteria from computational complexity theory.

Fiducial Inference

Fiducial inference was an approach to statistical inference based on fiducial probability, also known as a "fiducial distribution". In subsequent work, this approach has been called ill-defined, extremely limited in applicability, and even fallacious. However this argument is the same as that which shows that a so-called confidence distribution is not a valid probability distribution and, since this has not invalidated the application of confidence intervals, it does not necessarily invalidate conclusions drawn from fiducial arguments. An attempt was made to reinterpret the early work of Fisher's fiducial argument as a special case of an inference theory using Upper and lower probabilities.

Structural Inference

Developing ideas of Fisher and of Pitman from 1938 to 1939, George A. Barnard developed "structural inference" or "pivotal inference", an approach using invariant probabilities on group families. Barnard reformulated the arguments behind fiducial inference on a restricted class of models on which "fiducial" procedures would be well-defined and useful.

STATISTICAL ASSUMPTION

Statistics, like all mathematical disciplines, does not infer valid conclusions from nothing. Inferring interesting conclusions about real statistical populations almost always requires some background assumptions. Those assumptions must be made carefully, because incorrect assumptions can generate wildly inaccurate conclusions.

Here are some examples of statistical assumptions.

- Independence of observations from each other (this assumption is an especially common error).
- Independence of observational error from potential confounding effects.
- Exact or approximate normality of observations.
- Linearity of graded responses to quantitative stimuli, e.g. in linear regression.

Classes of Assumptions

There are two approaches to statistical inference: model-based inference and design-based inference. Both approaches rely on some statistical model to represent the data-generating process. In the model-based approach, the model is taken to be initially unknown, and one of the goals is to select an appropriate model for inference. In the design-based approach, the model is taken to be known, and one of the goals is to ensure that the sample data are selected randomly enough for inference.

Statistical assumptions can be put into two classes, depending upon which approach to inference is used.

- Model-based assumptions. These include the following three types:
  - Distributional assumptions. Where a statistical model involves terms relating to random errors, assumptions may be made about the probability distribution of these errors. In some cases, the distributional assumption relates to the observations themselves.
  - Structural assumptions. Statistical relationships between variables are often modelled by equating one variable to a function of another (or several others), plus a random error. Models often involve making a structural assumption about the form of the functional relationship, e.g. as in linear regression. This can be generalized to models involving relationships between underlying unobserved latent variables.
  - Cross-variation assumptions. These assumptions involve the joint probability distributions of either the observations themselves or the random errors in a model. Simple models may include the assumption that observations or errors are statistically independent.
- Design-based assumptions. These relate to the way observations have been gathered, and often involve an assumption of randomization during sampling.

The model-based approach is much the most commonly used in statistical inference; the design-based approach is used mainly with survey sampling. With the model-based approach, all the assumptions are effectively encoded in the model.

Checking Assumptions

Given that the validity of any conclusion drawn from a statistical inference depends on the validity of the assumptions made, it is clearly important that those assumptions should be reviewed at some stage. Some instances—for example where data are lacking—may require that researchers judge whether an assumption is reasonable. Researchers can expand this somewhat to consider what effect a departure from the assumptions might produce. Where more extensive data are available, various types of procedures for statistical model validation are available—e.g. for regression model validation.

ESTIMATION THEORY

Estimation theory is a branch of statistics that deals with estimating the values of parameters based on measured empirical data that has a random component. The parameters describe an underlying physical setting in such a way that their value affects the distribution of the measured data. An estimator attempts to approximate the unknown parameters using the measurements.

For example, it is desired to estimate the proportion of a population of voters who will vote for a particular candidate. That proportion is the parameter sought; the estimate is based on a small random sample of voters.

Or, for example, in radar the goal is to estimate the range of objects (airplanes, boats, etc.) by analyzing the two-way transit timing of received echoes of transmitted pulses. Since the reflected pulses are unavoidably embedded in electrical noise, their measured values are randomly distributed, so that the transit time must be estimated.

In estimation theory, two approaches are generally considered.

- The probabilistic approach assumes that the measured data is random with probability distribution dependent on the parameters of interest
- The set-membership approach assumes that the measured data vector belongs to a set which depends on the parameter vector.

For example, in electrical communication theory, the measurements which contain information regarding the parameters of interest are often associated with a noisy signal. Without randomness, or noise, the problem would be deterministic and estimation would not be needed.

Basics

To build a model, several statistical "ingredients" need to be known. These are needed to ensure the estimator has some mathematical tractability.

The first is a set of statistical samples taken from a random vector (RV) of size N. Put into a vector,

$$\mathbf{x} = \begin{bmatrix} x[0] \\ x[1] \\ \vdots \\ x[N-1] \end{bmatrix}.$$

Secondly, there are the corresponding M parameters

$$\boldsymbol{\theta} = \begin{bmatrix} \theta_1 \\ \theta_2 \\ \vdots \\ \theta_M \end{bmatrix},$$

which need to be established with their continuous probability density function (pdf) or its discrete counterpart, the probability mass function (pmf)

$$p(\mathbf{x}|\theta).$$

It is also possible for the parameters themselves to have a probability distribution (e.g., Bayesian statistics). It is then necessary to define the Bayesian probability

$$\pi(\theta).$$

After the model is formed, the goal is to estimate the parameters, commonly denoted $\hat{\theta}$, where the "hat" indicates the estimate.

One common estimator is the minimum mean squared error estimator, which utilizes the error between the estimated parameters and the actual value of the parameters

$$\mathbf{e} = \hat{\theta} - \theta$$

as the basis for optimality. This error term is then squared and minimized for the MMSE estimator.

STATISTICAL HYPOTHESIS TESTING

A statistical hypothesis is a hypothesis that is testable on the basis of observing a process that is modeled via a set of random variables. A statistical hypothesis test is a method of statistical inference. Commonly, two statistical data sets are compared, or a data set obtained by sampling is compared against a synthetic data set from an idealized model. A hypothesis is proposed for the statistical relationship between the two data sets, and this is compared as an alternative to an idealized null hypothesis that proposes no relationship between two data sets. The comparison is deemed statistically significant if the relationship between the data sets would be an unlikely realization of the null hypothesis according to a threshold probability—the significance level. Hypothesis tests are used in determining what outcomes of a study would lead to a rejection of the null hypothesis for a pre-specified level of significance. The process of distinguishing between the null hypothesis and the alternative hypothesis is aided by identifying two conceptual types of errors (type 1 & type 2), and by specifying parametric limits on e.g. how much type 1 error will be permitted.

An alternative framework for statistical hypothesis testing is to specify a set of statistical models, one for each candidate hypothesis, and then use model selection techniques to choose the most appropriate model. The most common selection techniques are based on either Akaike information criterion or Bayes factor.

Statistical hypothesis testing is sometimes called confirmatory data analysis. It can be contrasted with exploratory data analysis, which may not have pre-specified hypotheses.

Variations and Sub-Classes

Statistical hypothesis testing is a key technique of both frequentist inference and Bayesian inference, although the two types of inference have notable differences. Statistical hypothesis tests define a procedure that controls (fixes) the probability of incorrectly deciding that a default position (null hypothesis) is incorrect. The procedure is based on how likely it would be for a set of observations to occur if the null hypothesis were true. Note that this probability of making an incorrect decision is not the probability that the null hypothesis is true, nor whether any specific alternative hypothesis is true. This contrasts with other possible techniques of decision theory in which the null and alternative hypothesis are treated on a more equal basis.

One naïve Bayesian approach to hypothesis testing is to base decisions on the posterior probability, but this fails when comparing point and continuous hypotheses. Other approaches to decision making, such as Bayesian decision theory, attempt to balance the consequences of incorrect decisions across all possibilities, rather than concentrating on a single null hypothesis. A number of other approaches to reaching a decision based on data are available via decision theory and optimal decisions, some of which have desirable properties. Hypothesis testing, though, is a dominant approach to data analysis in many fields of science. Extensions to the theory of hypothesis testing include the study of the power of tests, i.e. the probability of correctly rejecting the null hypothesis given that it is false. Such considerations can be used for the purpose of sample size determination prior to the collection of data.

The Testing Process

In the statistics literature, statistical hypothesis testing plays a fundamental role. The usual line of reasoning is as follows:

- There is an initial research hypothesis of which the truth is unknown.
- The first step is to state the relevant null and alternative hypotheses. This is important, as mis-stating the hypotheses will muddy the rest of the process.
- The second step is to consider the statistical assumptions being made about the sample in doing the test; for example, assumptions about the statistical independence or about the form of the distributions of the observations. This is equally important as invalid assumptions will mean that the results of the test are invalid.
- Decide which test is appropriate, and state the relevant test statistic T.
- Derive the distribution of the test statistic under the null hypothesis from the assumptions. In standard cases this will be a well-known result. For example, the test statistic might follow a Student's t distribution or a normal distribution.
- Select a significance level (α), a probability threshold below which the null hypothesis will be rejected. Common values are 5% and 1%.
- The distribution of the test statistic under the null hypothesis partitions the possible values of T into those for which the null hypothesis is rejected—the so-called critical region—and those for which it is not. The probability of the critical region is α.
- Compute from the observations the observed value t_{obs} of the test statistic T.
- Decide to either reject the null hypothesis in favor of the alternative or not reject it. The decision rule is to reject the null hypothesis H_0 if the observed

value t_{obs} is in the critical region, and to accept or "fail to reject" the hypothesis otherwise.

An alternative process is commonly used:

- Compute from the observations the observed value t_{obs} of the test statistic T.
- Calculate the p-value. This is the probability, under the null hypothesis, of sampling a test statistic at least as extreme as that which was observed.
- Reject the null hypothesis, in favor of the alternative hypothesis, if and only if the p-value is less than the significance level (the selected probability) threshold.

The two processes are equivalent. The former process was advantageous in the past when only tables of test statistics at common probability thresholds were available. It allowed a decision to be made without the calculation of a probability. It was adequate for classwork and for operational use, but it was deficient for reporting results.

The latter process relied on extensive tables or on computational support not always available. The explicit calculation of a probability is useful for reporting. The calculations are now trivially performed with appropriate software.

The difference in the two processes applied to the Radioactive suitcase example (below):

- "The Geiger-counter reading is 10. The limit is 9. Check the suitcase."
- "The Geiger-counter reading is high; 97% of safe suitcases have lower readings. The limit is 95%. Check the suitcase."

The former report is adequate, the latter gives a more detailed explanation of the data and the reason why the suitcase is being checked.

It is important to note the difference between accepting the null hypothesis and simply failing to reject it. The "fail to reject" terminology highlights the fact that the null hypothesis is assumed to be true from the start of the test; if there is a lack of evidence against it, it simply continues to be assumed true. The phrase "accept the null hypothesis" may suggest it has been proved simply because it has not been disproved, a logical fallacy known as the argument from ignorance. Unless a test with particularly high power is used, the idea of "accepting" the null hypothesis may be dangerous. Nonetheless the terminology is prevalent throughout statistics, where the meaning actually intended is well understood.

The processes described here are perfectly adequate for computation. They seriously neglect the design of experiments considerations.

It is particularly critical that appropriate sample sizes be estimated before conducting the experiment.

The phrase "test of significance" was coined by statistician Ronald Fisher.

Interpretation

If the p-value is less than the required significance level (equivalently, if the observed test statistic is in the critical region), then we say the null hypothesis is rejected at the given level of significance. Rejection of the null hypothesis is a conclusion. This is like a "guilty" verdict in a criminal trial: the evidence is sufficient to reject innocence, thus proving guilt. We might accept the alternative hypothesis (and the research hypothesis).

If the p-value is not less than the required significance level (equivalently, if the observed test statistic is outside the critical region), then the test has no result. The evidence is insufficient to support a conclusion. (This is like a jury that fails to reach a verdict.) The researcher typically gives extra consideration to those cases where the p-value is close to the significance level.

In the Lady tasting tea example (below), Fisher required the Lady to properly categorize all of the cups of tea to justify the conclusion that the result was unlikely to result from chance. He defined the critical region as that case alone. The region was defined by a probability (that the null hypothesis was correct) of less than 5%.

Whether rejection of the null hypothesis truly justifies acceptance of the research hypothesis depends on the structure of the hypotheses. Rejecting the hypothesis that a large paw print originated from a bear does not immediately prove the existence of Bigfoot. Hypothesis testing emphasizes the rejection, which is based on a probability, rather than the acceptance, which requires extra steps of logic.

"The probability of rejecting the null hypothesis is a function of five factors: whether the test is one- or two tailed, the level of significance, the standard deviation, the amount of deviation from the null hypothesis, and the number of observations." These factors are a source of criticism; factors under the control of the experimenter/analyst give the results an appearance of subjectivity.

Use and Importance

Statistics are helpful in analyzing most collections of data. This is equally true of hypothesis testing which can justify conclusions even when no scientific theory exists. In the Lady tasting tea example, it was "obvious" that no difference existed between (milk poured into tea) and (tea poured into milk). The data contradicted the "obvious".

Real world applications of hypothesis testing include:

- Testing whether more men than women suffer from nightmares
- Establishing authorship of documents
- Evaluating the effect of the full moon on behavior
- Determining the range at which a bat can detect an insect by echo
- Deciding whether hospital carpeting results in more infections
- Selecting the best means to stop smoking
- Checking whether bumper stickers reflect car owner behavior
- Testing the claims of handwriting analysts

Statistical hypothesis testing plays an important role in the whole of statistics and in statistical inference. For example, Lehmann (1992) in a review of the fundamental paper by Neyman and Pearson (1933) says: "Nevertheless, despite their shortcomings, the new paradigm formulated in the 1933 paper, and the many developments carried out within its framework continue to play a central role in both the theory and practice of statistics and can be expected to do so in the foreseeable future". Significance testing has been the favored statistical tool in some experimental social sciences (over 90% of articles in the Journal of Applied Psychology during the early 1990s). Other fields have favored the estimation of parameters (e.g., effect size). Significance testing is used as a substitute for the traditional comparison of predicted value and experimental result at the core of the scientific method. When theory is only capable of predicting the sign of a relationship, a directional (one-sided) hypothesis test can be configured so that only a statistically significant result supports theory. This form of theory appraisal is the most heavily criticized application of hypothesis testing.

Cautions

"If the government required statistical procedures to carry warning labels like those on drugs, most inference methods would have long labels indeed." This caution applies to hypothesis tests and alternatives to them.

The successful hypothesis test is associated with a probability and a type-I error rate. The conclusion might be wrong.

The conclusion of the test is only as solid as the sample upon which it is based. The design of the experiment is critical. A number of unexpected effects have been observed including:

- The clever Hans effect. A horse appeared to be capable of doing simple arithmetic.
- The Hawthorne effect. Industrial workers were more productive in better illumination, and most productive in worse.
- The placebo effect. Pills with no medically active ingredients were remarkably effective.

A statistical analysis of misleading data produces misleading conclusions. The issue of data quality can be more subtle. In forecasting for example, there is no agreement on a measure of forecast accuracy. In the absence of a consensus measurement, no decision based on measurements will be without controversy.

The book How to Lie with Statistics is the most popular book on statistics ever published. It does not much consider hypothesis testing, but its cautions are applicable, including: Many claims are made on the basis of samples too small to convince. If a report does not mention sample size, be doubtful.

Hypothesis testing acts as a filter of statistical conclusions; only those results meeting a probability threshold are publishable. Economics also acts as a publication filter; only those results favorable to the author and funding source may be submitted for publication. The impact of filtering on publication is termed publication bias. A related problem is that of multiple testing (sometimes linked to data mining), in which a variety of tests for a variety of possible effects are applied to a single data set and only those yielding a significant result are reported. These are often dealt with by using multiplicity correction procedures that control the family wise error rate (FWER) or the false discovery rate (FDR). Those making critical decisions based on the results of a hypothesis test are prudent to look at the details rather than the conclusion alone. In the physical sciences most results are fully accepted only when independently confirmed. The general advice concerning statistics is, “Figures never lie, but liars figure” (anonymous).

Examples

Lady Tasting Tea

In a famous example of hypothesis testing, known as the Lady tasting tea, a female colleague of Fisher claimed to be able to tell whether the tea or the milk was added first to a cup. Fisher proposed to give her eight cups, four of each variety, in random order. One could then ask what the probability was for her getting the number she got correct, but just by chance. The null hypothesis was that the Lady had no such ability. The test statistic was a simple count of the number of successes in selecting the 4 cups. The critical region was the single case of 4 successes of 4 possible based on a conventional probability criterion (< 5%; 1 of 70 ≈ 1.4%). Fisher asserted that no alternative hypothesis was (ever) required. The lady correctly identified every cup, which would be considered a statistically significant result.

Courtroom Trial

A statistical test procedure is comparable to a criminal trial; a defendant is considered not guilty as long as his or her guilt is not proven. The prosecutor tries to prove the guilt of the defendant. Only when there is enough charging evidence the defendant is convicted.

In the start of the procedure, there are two hypotheses H_0: “the defendant is not guilty”, and H_1: “the defendant is guilty”. The first one is called null hypothesis, and is for the time being accepted. The second one is called alternative (hypothesis). It is the hypothesis one hopes to support.

The hypothesis of innocence is only rejected when an error is very unlikely, because one doesn’t want to convict an innocent defendant. Such an error is called error of the first kind (i.e., the conviction of an innocent person), and the occurrence of this error is controlled to be rare. As a consequence of this asymmetric behavior, the error of the second kind (acquitting a person who committed the crime), is often rather large.

	H_0 is true **Truly not guilty**	**H_1 is true** **Truly guilty**
Accept null hypothesis Acquittal	Right decision	Wrong decision Type II Error
Reject null hypothesis Conviction	Wrong decision Type I Error	Right decision

A criminal trial can be regarded as either or both of two decision processes: guilty vs not guilty or evidence vs a threshold (“beyond a reasonable doubt”). In one view, the defendant is judged; in the other view the performance of the prosecution (which bears the burden of proof) is judged. A hypothesis test can be regarded as either a judgment of a hypothesis or as a judgment of evidence.

Philosopher's Beans

The following example was produced by a philosopher describing scientific methods generations before hypothesis testing was formalized and popularized.

Few beans of this handful are white.

Most beans in this bag are white.

Therefore: Probably, these beans were taken from another bag.

This is an hypothetical inference.

The beans in the bag are the population. The handful are the sample. The null hypothesis is that the sample originated from the population. The criterion for rejecting the null-hypothesis is the "obvious" difference in appearance (an informal difference in the mean). The interesting result is that consideration of a real population and a real sample produced an imaginary bag. The philosopher was considering logic rather than probability. To be a real statistical hypothesis test, this example requires the formalities of a probability calculation and a comparison of that probability to a standard.

A simple generalization of the example considers a mixed bag of beans and a handful that contain either very few or very many white beans. The generalization considers both extremes. It requires more calculations and more comparisons to arrive at a formal answer, but the core philosophy is unchanged; If the composition of the handful is greatly different from that of the bag, then the sample probably originated from another bag. The original example is termed a one-sided or a one-tailed test while the generalization is termed a two-sided or two-tailed test.

The statement also relies on the inference that the sampling was random. If someone had been picking through the bag to find white beans, then it would explain why the handful had so many white beans, and also explain why the number of white beans in the bag was depleted (although the bag is probably intended to be assumed much larger than one's hand).

Clairvoyant Card Game

A person (the subject) is tested for clairvoyance. He is shown the reverse of a randomly chosen playing card 25 times and asked which of the four suits it belongs to. The number of hits, or correct answers, is called X.

As we try to find evidence of his clairvoyance, for the time being the null hypothesis is that the person is not clairvoyant. The alternative is, of course: the person is (more or less) clairvoyant.

If the null hypothesis is valid, the only thing the test person can do is guess. For every card, the probability (relative frequency) of any single suit appearing is 1/4. If the alternative is valid, the test subject will predict the suit correctly with probability greater than 1/4. We will call the probability of guessing correctly p. The hypotheses, then, are:

- null hypothesis $H_0 : p = \frac{1}{4}$ (just guessing)

  and

- alternative hypothesis $H_1 : p > \frac{1}{4}$ (true clairvoyant).

When the test subject correctly predicts all 25 cards, we will consider him clairvoyant, and reject the null hypothesis. Thus also with 24 or 23 hits. With only 5 or 6 hits, on the other hand, there is no cause to consider him so. But what about 12 hits, or 17 hits? What is the critical number, c, of hits, at which point we consider the subject to be clairvoyant? How do we determine the critical value c? It is obvious that with the choice c=25 (i.e. we only accept clairvoyance when all cards are predicted correctly) we're more critical than with c=10. In the first case almost no test subjects will be recognized to be clairvoyant, in the second case, a certain number will pass the test. In practice, one decides how critical one will be. That is, one decides how often one accepts an error of the first kind – a false positive, or Type I error. With c = 25 the probability of such an error is:

$$P(\text{reject } H_0 \mid H_0 \text{ is valid}) = P(X = 25 \mid p = \tfrac{1}{4}) = \left(\tfrac{1}{4}\right)^{25} \approx 10^{-15},$$

and hence, very small. The probability of a false positive is the probability of randomly guessing correctly all 25 times.

Being less critical, with c=10, gives:

$$P(\text{reject } H_0 \mid H_0 \text{ is valid}) = P(X \geq 10 | p = \tfrac{1}{4}) = \sum_{k=10}^{25} P(X = k \mid p = \tfrac{1}{4}) \approx 0.07.$$

Thus, c = 10 yields a much greater probability of false positive.

Before the test is actually performed, the maximum acceptable probability of a Type I error (α) is determined. Typically, values in the range of 1% to 5% are selected. (If the maximum acceptable error rate is zero, an infinite number of correct guesses is required.) Depending on this Type 1 error rate, the critical value c is calculated. For example, if we select an error rate of 1%, c is calculated thus:

$$P(\text{reject } H_0 \mid H_0 \text{ is valid}) = P(X \geq c \mid p = \tfrac{1}{4}) \leq 0.01.$$

From all the numbers c, with this property, we choose the smallest, in order to minimize the probability of a Type II error, a false negative. For the above example, we select: c=13.

Radioactive Suitcase

As an example, consider determining whether a suitcase contains some radioactive material. Placed under a Geiger counter, it produces 10 counts per minute. The null hypothesis is that no radioactive material is in the suitcase and that all measured counts are due to ambient radioactivity typical of the surrounding air and harmless objects. We can then calculate how likely it is that we would observe 10 counts per minute if the null hypothesis were true. If the null hypothesis predicts (say) on average 9 counts per minute, then according to the Poisson distribution typical for radioactive decay there is about 41% chance of recording 10 or more counts. Thus we can say that the suitcase is compatible with the null hypothesis (this does not guarantee that there is no radioactive material, just that we don't have enough evidence to suggest there is). On the other hand, if the null hypothesis predicts 3 counts per minute (for which the Poisson distribution predicts only 0.1% chance of recording 10 or more counts) then the suitcase is not compatible with the null hypothesis, and there are likely other factors responsible to produce the measurements.

The test does not directly assert the presence of radioactive material. A successful test asserts that the claim of no radioactive material present is unlikely given the reading (and therefore ...). The double negative (disproving the null hypothesis) of the method is confusing, but using a counter-example to disprove is standard mathematical practice. The attraction of the method is its practicality. We know (from experience) the expected range of counts with only ambient radioactivity present, so we can say that a measurement is unusually large. Statistics just formalizes the intuitive by using numbers instead of adjectives. We probably do not know the characteristics of the radioactive suitcases; We just assume that they produce larger readings.

To slightly formalize intuition: Radioactivity is suspected if the Geiger-count with the suitcase is among or exceeds the greatest (5% or 1%) of the Geiger-counts made with ambient radiation alone. This makes no assumptions about the distribution of counts. Many ambient radiation observations are required to obtain good probability estimates for rare events.

The test described here is more fully the null-hypothesis statistical significance test. The null hypothesis represents what we would believe by default, before seeing any evidence. Statistical significance is a possible finding of the test, declared when the observed sample is unlikely to have occurred by chance if the null hypothesis were true. The name of the test describes its formulation and its possible outcome. One characteristic of the test is its crisp decision: to reject or not reject the null hypothesis. A calculated value is compared to a threshold, which is determined from the tolerable risk of error.

Definition of Terms

The following definitions are mainly based on the exposition in the book by Lehmann and Romano:

Statistical Hypothesis

A statement about the parameters describing a population (not a sample).

Statistic

A value calculated from a sample, often to summarize the sample for comparison purposes.

Simple Hypothesis

Any hypothesis which specifies the population distribution completely.

Composite Hypothesis

Any hypothesis which does not specify the population distribution completely.

Null Hypothesis (H_0)

A simple hypothesis associated with a contradiction to a theory one would like to prove.

Alternative Hypothesis (H_1)

A hypothesis (often composite) associated with a theory one would like to prove.

Statistical Test

A procedure whose inputs are samples and whose result is a hypothesis.

Region of Acceptance

The set of values of the test statistic for which we fail to reject the null hypothesis.

Region of Rejection / Critical Region

The set of values of the test statistic for which the null hypothesis is rejected.

Critical Value

The threshold value delimiting the regions of acceptance and rejection for the test statistic.

Power of a Test ($1 - \beta$)

The test's probability of correctly rejecting the null hypothesis. The complement of the false negative rate, β. Power is termed sensitivity in biostatistics. (This is a sensitive test. Because the result is negative, we can confidently say that the patient does not have the condition.)

Size

For simple hypotheses, this is the test's probability of incorrectly rejecting the null hypothesis. The false positive rate. For composite hypotheses this is the supremum of the probability of rejecting the null hypothesis over all cases covered by the null hypothesis. The complement of the false positive rate is termed specificity in biostatistics. ("This is a specific test. Because the result is positive, we can confidently say that the patient has the condition.")

Significance Level of a Test (α)

It is the upper bound imposed on the size of a test. Its value is chosen by the statistician prior to looking at the data or choosing any particular test to be used. It is the maximum exposure to erroneously rejecting H_0 he/she is ready to accept. Testing H_0 at significance level α means testing H_0 with a test whose size does not exceed α. In most cases, one uses tests whose size is equal to the significance level.

p-value

The probability, assuming the null hypothesis is true, of observing a result at least as extreme as the test statistic.

Statistical Significance Test

A predecessor to the statistical hypothesis test. An experimental result was said to be statistically significant if a sample was sufficiently inconsistent with the (null) hypothesis. This was variously considered common sense, a pragmatic heuristic for identifying meaningful experimental results, a convention establishing a threshold of statistical evidence or a method for drawing conclusions from data. The statistical hypothesis test added mathematical rigor and philosophical consistency to the concept by making the alternative hypothesis explicit. The term is loosely used to describe the modern version which is now part of statistical hypothesis testing.

Conservative Test

A test is conservative if, when constructed for a given nominal significance level, the true probability of incorrectly rejecting the null hypothesis is never greater than the nominal level.

Exact Test

A test in which the significance level or critical value can be computed exactly, i.e., without any approximation. In some contexts this term is restricted to tests applied to categorical data and to permutation tests, in which computations are carried out by complete enumeration of all possible outcomes and their probabilities.

A statistical hypothesis test compares a test statistic (z or t for examples) to a threshold. The test statistic (the formula found in the table below) is based on optimality. For a fixed level of Type I error rate, use of these statistics minimizes Type II error rates (equivalent to maximizing power). The following terms describe tests in terms of such optimality:

Most Powerful Test

For a given size or significance level, the test with the greatest power (probability of rejection) for a given value of the parameter(s) being tested, contained in the alternative hypothesis.

Uniformly Most Powerful Test (UMP)

A test with the greatest power for all values of the parameter(s) being tested, contained in the alternative hypothesis.

Common Test Statistics

One-sample tests are appropriate when a sample is being compared to the population from a hypothesis. The population characteristics are known from theory or are calculated from the population.

Two-sample tests are appropriate for comparing two samples, typically experimental and control samples from a scientifically controlled experiment.

Paired tests are appropriate for comparing two samples where it is impossible to control important variables. Rather than comparing two sets, members are paired between samples so the difference between the members becomes the sample. Typically the mean of the differences is then compared to zero. The common example scenario for when a paired difference test is appropriate is when a single set of test subjects has something applied to them and the test is intended to check for an effect.

Z-tests are appropriate for comparing means under stringent conditions regarding normality and a known standard deviation.

A t-test is appropriate for comparing means under relaxed conditions (less is assumed).

Tests of proportions are analogous to tests of means (the 50% proportion).

Chi-squared tests use the same calculations and the same probability distribution for different applications:

- Chi-squared tests for variance are used to determine whether a normal population has a specified variance. The null hypothesis is that it does.
- Chi-squared tests of independence are used for deciding whether two variables are associated or are independent. The variables are categorical rather than numeric. It can be used to decide whether left-handedness is correlated with libertarian politics (or not). The null hypothesis is that the variables are independent. The numbers used in the calculation are the observed and expected frequencies of occurrence (from contingency tables).
- Chi-squared goodness of fit tests are used to determine the adequacy of curves fit to data. The null hypothesis is that the curve fit is adequate. It is common to determine curve shapes to minimize the mean square error, so it is appropriate that the goodness-of-fit calculation sums the squared errors.

F-tests (analysis of variance, ANOVA) are commonly used when deciding whether groupings of data by category are meaningful. If the variance of test scores of the left-handed in a class is much smaller than the variance of the whole class, then it may be useful to study lefties as a group. The null hypothesis is that two variances are the same – so the proposed grouping is not meaningful.

In the table below, the symbols used are defined at the bottom of the table. Proofs exist that the test statistics are appropriate.

Name	Formula	Assumptions or notes
One-sample z-test	$z = \frac{\overline{x} - \mu_0}{(\sigma/\sqrt{n})}$	(Normal population or $n > 30$) and σ known. (z is the distance from the mean in relation to the standard deviation of the mean). For non-normal distributions it is possible to calculate a minimum proportion of a population that falls within k standard deviations for any k (see: Chebyshev's inequality).
Two-sample z-test	$z = \frac{(\overline{x}_1 - \overline{x}_2) - d_0}{\sqrt{\frac{\sigma_1^2}{n_1} + \frac{\sigma_2^2}{n_2}}}$	Normal population and independent observations and σ_1 and σ_2 are known
One-sample t-test	$t = \frac{\overline{x} - \mu_0}{(s/\sqrt{n})}$, $df = n - 1$	(Normal population or $n < 30$) and σ unknown
Paired t-test	$t = \frac{\overline{d} - d_0}{(s_d/\sqrt{n})}$, $df = n - 1$	(Normal population of differences or $n < 30$) and σ unknown or small sample size $n < 30$
Two-sample pooled t-test, equal variances	$t = \frac{(\overline{x}_1 - \overline{x}_2) - d_0}{s_p\sqrt{\frac{1}{n_1} + \frac{1}{n_2}}}$, $s_p^2 = \frac{(n_1 - 1)s_1^2 + (n_2 - 1)s_2^2}{n_1 + n_2 - 2}$, $df = n_1 + n_2 - 2$ [22]	(Normal populations or $n_1 + n_2 > 40$) and independent observations and $\sigma_1 = \sigma_2$ unknown

Two-sample unpooled *t*-test, unequal variances (Welch's *t*-test)	$t = \dfrac{(\bar{x}_1 - \bar{x}_2) - d_0}{\sqrt{\frac{s_1^2}{n_1} + \frac{s_2^2}{n_2}}}$, $df = \dfrac{\left(\frac{s_1^2}{n_1} + \frac{s_2^2}{n_2}\right)^2}{\frac{\left(\frac{s_1^2}{n_1}\right)^2}{n_1 - 1} + \frac{\left(\frac{s_2^2}{n_2}\right)^2}{n_2 - 1}}$	(Normal populations **or** $n_1 + n_2 > 40$) **and** independent observations **and** $\sigma_1 \neq \sigma_2$ both unknown
One-proportion z-test	$z = \dfrac{\hat{p} - p_0}{\sqrt{p_0(1 - p_0)}}\sqrt{n}$	$n\ p_0 > 10$ **and** $n\ (1 - p_0) > 10$ **and** it is a SRS (Simple Random Sample)
Two-proportion z-test, pooled for $H_0\colon p_1 = p_2$	$z = \dfrac{(\hat{p}_1 - \hat{p}_2)}{\sqrt{\hat{p}(1 - \hat{p})(\frac{1}{n_1} + \frac{1}{n_2})}}$ $\hat{p} = \dfrac{x_1 + x_2}{n_1 + n_2}$	$n_1\ p_1 > 5$ **and** $n_1(1 - p_1) > 5$ **and** $n_2\ p_2 > 5$ **and** $n_2(1 - p_2) > 5$ **and** independent observations
Two-proportion z-test, unpooled for $\lvert d_0 \rvert > 0$	$z = \dfrac{(\hat{p}_1 - \hat{p}_2) - d_0}{\sqrt{\frac{\hat{p}_1(1 - \hat{p}_1)}{n_1} + \frac{\hat{p}_2(1 - \hat{p}_2)}{n_2}}}$	$n_1\ p_1 > 5$ **and** $n_1(1 - p_1) > 5$ **and** $n_2\ p_2 > 5$ **and** $n_2(1 - p_2) > 5$ **and** independent observations
Chi-squared test for variance	$\chi^2 = (n - 1)\dfrac{s^2}{\sigma_0^2}$	Normal population
Chi-squared test for goodness of fit	$\chi^2 = \sum^{k} \dfrac{(\text{observed} - \text{expected})^2}{\text{expected}}$	*df* = *k* - 1 - *# parameters estimated*, and one of these must hold. • All expected counts are at least 5. • All expected counts are > 1 and no more than 20% of expected counts are less than 5

Two-sample F test for equality of variances	$F = \dfrac{s_1^2}{s_2^2}$	Normal populations Arrange so $s_1^2 \geq s_2^2$ and reject H_0 for $F > F(\alpha/2, n_1 - 1, n_2 - 1)$
Regression *t*-test of $H_0\colon R^2 = 0.$	$t = \sqrt{\dfrac{R^2(n - k - 1^*)}{1 - R^2}}$	Reject H_0 for $t > t(\alpha/2, n - k - 1^*)$ *Subtract 1 for intercept; *k* terms contain independent variables.

In general, the subscript 0 indicates a value taken from the null hypothesis, H_0, which should be used as much as possible in constructing its test statistic. ... *Definitions of other symbols:*

- α, the probability of Type I error (rejecting a null hypothesis when it is in fact true)
- n = sample size
- n_1 = sample 1 size
- n_2 = sample 2 size
- $\bar{x}$ = sample mean
- μ_0 = hypothesized population mean
- μ_1 = population 1 mean
- μ_2 = population 2 mean
- σ = population standard deviation
- σ^2 = population variance
- s = sample standard deviation
- $\sum^{k}$ = sum (of *k* numbers)
- s^2 = sample variance
- s_1 = sample 1 standard deviation
- s_2 = sample 2 standard deviation
- t = t statistic
- df = degrees of freedom
- $\bar{d}$ = sample mean of differences
- d_0 = hypothesized population mean difference
- s_d = standard deviation of differences
- χ^2 = Chi-squared statistic
- $\hat{p}$ = *x/n* = sample proportion, unless specified otherwise
- p_0 = hypothesized population proportion
- p_1 = proportion 1
- p_2 = proportion 2
- d_p = hypothesized difference in proportion
- $\min\{n_1, n_2\}$ = minimum of n_1 and n_2
- $x_1 = n_1 p_1$
- $x_2 = n_2 p_2$
- F = F statistic

Origins and Early Controversy

Significance testing is largely the product of Karl Pearson (p-value, Pearson's chi-squared test), William Sealy Gosset (Student's t-distribution), and Ronald Fisher ("null hypothesis", analysis of variance, "significance test"), while hypothesis testing was developed by Jerzy

Neyman and Egon Pearson (son of Karl). Ronald Fisher began his life in statistics as a Bayesian (Zabell 1992), but Fisher soon grew disenchanted with the subjectivity involved (namely use of the principle of indifference when determining prior probabilities), and sought to provide a more "objective" approach to inductive inference.

Fisher was an agricultural statistician who emphasized rigorous experimental design and methods to extract a result from few samples assuming Gaussian distributions. Neyman (who teamed with the younger Pearson) emphasized mathematical rigor and methods to obtain more results from many samples and a wider range of distributions. Modern hypothesis testing is an inconsistent hybrid of the Fisher vs Neyman/Pearson formulation, methods and terminology developed in the early 20th century. While hypothesis testing was popularized early in the 20th century, evidence of its use can be found much earlier. In the 1770s Laplace considered the statistics of almost half a million births. The statistics showed an excess of boys compared to girls. He concluded by calculation of a p-value that the excess was a real, but unexplained, effect.

Fisher popularized the "significance test". He required a null-hypothesis (corresponding to a population frequency distribution) and a sample. His (now familiar) calculations determined whether to reject the null-hypothesis or not. Significance testing did not utilize an alternative hypothesis so there was no concept of a Type II error.

The p-value was devised as an informal, but objective, index meant to help a researcher determine (based on other knowledge) whether to modify future experiments or strengthen one's faith in the null hypothesis. Hypothesis testing (and Type I/II errors) was devised by Neyman and Pearson as a more objective alternative to Fisher's p-value, also meant to determine researcher behavior, but without requiring any inductive inference by the researcher.

Neyman & Pearson considered a different problem (which they called "hypothesis testing"). They initially considered two simple hypotheses (both with frequency distributions). They calculated two probabilities and typically selected the hypothesis associated with the higher probability (the hypothesis more likely to have generated the sample). Their method always selected a hypothesis. It also allowed the calculation of both types of error probabilities.

Fisher and Neyman/Pearson clashed bitterly. Neyman/Pearson considered their formulation to be an improved generalization of significance testing. (The defining paper was abstract. Mathematicians have generalized and refined the theory for decades.) Fisher thought that it was not applicable to scientific research because often, during the course of the experiment, it is discovered that the initial assumptions about the null hypothesis are questionable due to unexpected sources of error. He believed that the use of rigid reject/accept decisions based on models formulated before data is collected was incompatible with this common scenario faced by scientists and attempts to apply this method to scientific research would lead to mass confusion.

The dispute between Fisher and Neyman–Pearson was waged on philosophical grounds, characterized by a philosopher as a dispute over the proper role of models in statistical inference.

Events intervened: Neyman accepted a position in the western hemisphere, breaking his partnership with Pearson and separating disputants (who had occupied the same building) by much of the planetary diameter. World War II provided an intermission in the debate. The dispute between Fisher and Neyman terminated (unresolved after 27 years) with Fisher's death in 1962. Neyman wrote a well-regarded eulogy. Some of Neyman's later publications reported p-values and significance levels.

The modern version of hypothesis testing is a hybrid of the two approaches that resulted from confusion by writers of statistical textbooks (as predicted by Fisher) beginning in the 1940s. (But signal detection, for example, still uses the Neyman/Pearson formulation.) Great conceptual differences and many caveats in addition to those mentioned above were ignored. Neyman and Pearson provided the stronger terminology, the more rigorous mathematics and the more consistent philosophy, but the subject taught today in introductory statistics has more similarities with Fisher's method than theirs. This history explains the inconsistent terminology (example: the null hypothesis is never accepted, but there is a region of acceptance).

Sometime around 1940, in an apparent effort to provide researchers with a "non-controversial" way to have their cake and eat it too, the authors of statistical text books began anonymously combining these two strategies by using the p-value in place of the test statistic (or data) to test against the Neyman–Pearson "significance level". Thus, researchers were encouraged to infer the strength of their data against some null hypothesis using p-values, while also thinking they are retaining the post-data collection objectivity provided by hypothesis testing. It then became customary for the null hypothesis, which was originally some realistic research hypothesis, to be used almost solely as a strawman "nil" hypothesis (one where a treatment has no effect, regardless of the context).

A comparison between Fisherian, frequentist (Neyman–Pearson)

Fisher's null hypothesis testing	**Neyman–Pearson decision theory**
1. Set up a statistical null hypothesis. The null need not be a nil hypothesis (i.e., zero difference).	1. Set up two statistical hypotheses, H1 and H2, and decide about α, β, and sample size before the experiment, based on subjective cost-benefit considerations. These define a rejection region for each hypothesis.
2. Report the exact level of significance (e.g., $p = 0.051$ or $p = 0.049$). Do not use a conventional 5% level, and do not talk about accepting or rejecting hypotheses. If the result is "not significant", draw no conclusions and make no decisions, but suspend judgement until further data is available.	2. If the data falls into the rejection region of H1, accept H2; otherwise accept H1. Note that accepting a hypothesis does not mean that you believe in it, but only that you act as if it were true.
3. Use this procedure only if little is known about the problem at hand, and only to draw provisional conclusions in the context of an attempt to understand the experimental situation.	3. The usefulness of the procedure is limited among others to situations where you have a disjunction of hypotheses (e.g., either $\mu1 = 8$ or $\mu2 = 10$ is true) and where you can make meaningful cost-benefit trade-offs for choosing alpha and beta.

Early Choices of Null Hypothesis

Paul Meehl has argued that the epistemological importance of the choice of null hypothesis has gone largely unacknowledged. When the null hypothesis is predicted by theory, a more precise experiment will be a more severe test of the underlying theory. When the null hypothesis defaults to "no difference" or "no effect", a more precise experiment is a less severe test of the theory that motivated performing the experiment. An examination of the origins of the latter practice may therefore be useful:

1778: Pierre Laplace compares the birthrates of boys and girls in multiple European cities. He states: "it is natural to conclude that these possibilities are very nearly in the same ratio". Thus Laplace's null hypothesis that the birthrates of boys and girls should be equal given "conventional wisdom".

1900: Karl Pearson develops the chi squared test to determine "whether a given form of frequency curve will effectively describe the samples drawn from a given population." Thus the null hypothesis is that a population is described by some distribution predicted by theory. He uses as an example the numbers of five and sixes in the Weldon dice throw data.

1904: Karl Pearson develops the concept of "contingency" in order to determine whether outcomes are independent of a given categorical factor. Here the null hypothesis is by default that two things are unrelated (e.g. scar formation and death rates from smallpox). The null hypothesis in this case is no longer predicted by theory or conventional wisdom, but is instead the principle of indifference that lead Fisher and others to dismiss the use of "inverse probabilities".

Versus Null Hypothesis Statistical Significance Testing

An example of Neyman–Pearson hypothesis testing can be made by a change to the radioactive suitcase example. If the "suitcase" is actually a shielded container for the transportation of radioactive material, then a test might be used to select among three hypotheses: no radioactive source present, one present, two (all) present. The test could be required for safety, with actions required in each case. The Neyman–Pearson lemma of hypothesis testing says that a good criterion for the selection of hypotheses is the ratio of their probabilities (a likelihood ratio). A simple method of solution is to select the hypothesis with the highest probability for the Geiger counts observed. The typical result matches intuition: few counts imply no source, many counts imply two sources and intermediate counts imply one source.

Neyman–Pearson theory can accommodate both prior probabilities and the costs of actions resulting from decisions. The former allows each test to consider the results of earlier tests (unlike Fisher's significance tests). The latter allows the consideration of economic issues (for example) as well as probabilities. A likelihood ratio remains a good criterion for selecting among hypotheses.

The two forms of hypothesis testing are based on different problem formulations. The original test is analogous to a true/false question; the Neyman–Pearson test is more like multiple choice. In the view of Tukey the former produces a conclusion on the basis of only strong evidence while the latter produces a decision on the basis of available evidence. While the two tests seem quite different both mathematically and philosophically, later developments lead to the opposite claim. Consider many tiny radioactive sources.

The hypotheses become 0,1,2,3... grains of radioactive sand. There is little distinction between none or some radiation (Fisher) and 0 grains of radioactive sand versus all of the alternatives (Neyman–Pearson). The major Neyman–Pearson paper of 1933 also considered composite hypotheses (ones whose distribution includes an unknown parameter). An example proved the optimality of the (Student's) t-test, "there can be no better test for the hypothesis under consideration" (p 321). Neyman–Pearson theory was proving the optimality of Fisherian methods from its inception.

Fisher's significance testing has proven a popular flexible statistical tool in application with little mathematical growth potential. Neyman–Pearson hypothesis testing is claimed as a pillar of mathematical statistics, creating a new paradigm for the field. It also stimulated new applications in Statistical process control, detection theory, decision theory and game theory. Both formulations have been successful, but the successes have been of a different character.

The dispute over formulations is unresolved. Science primarily uses Fisher's (slightly modified) formulation as taught in introductory statistics. Statisticians study Neyman–Pearson theory in graduate school. Mathematicians are proud of uniting the formulations. Philosophers consider them separately. Learned opinions deem the formulations variously competitive (Fisher vs Neyman), incompatible or complementary. The dispute has become more complex since Bayesian inference has achieved respectability.

The terminology is inconsistent. Hypothesis testing can mean any mixture of two formulations that both changed with time. Any discussion of significance testing vs hypothesis testing is doubly vulnerable to confusion.

Fisher thought that hypothesis testing was a useful strategy for performing industrial quality control, however, he strongly disagreed that hypothesis testing could be useful for scientists. Hypothesis testing provides a means of finding test statistics used in significance testing. The concept of power is useful in explaining the consequences of adjusting the significance level and is heavily used in sample size determination. The two methods remain philosophically distinct. They usually (but not always) produce the same mathematical answer. The preferred answer is context dependent. While the existing merger of Fisher and Neyman–Pearson theories has been heavily criticized, modifying the merger to achieve Bayesian goals has been considered.

Criticism

Criticism of statistical hypothesis testing fills volumes citing 300–400 primary references. Much of the criticism can be summarized by the following issues:

- The interpretation of a p-value is dependent upon stopping rule and definition of multiple comparison. The former often changes during the course of a study and the latter is unavoidably ambiguous. (i.e. "p values depend on both the (data) observed and on the other possible (data) that might have been observed but weren't").
- Confusion resulting (in part) from combining the methods of Fisher and Neyman–Pearson which are conceptually distinct.
- Emphasis on statistical significance to the exclusion of estimation and confirmation

by repeated experiments.

- Rigidly requiring statistical significance as a criterion for publication, resulting in publication bias. Most of the criticism is indirect. Rather than being wrong, statistical hypothesis testing is misunderstood, overused and misused.
- When used to detect whether a difference exists between groups, a paradox arises. As improvements are made to experimental design (e.g., increased precision of measurement and sample size), the test becomes more lenient. Unless one accepts the absurd assumption that all sources of noise in the data cancel out completely, the chance of finding statistical significance in either direction approaches 100%.
- Layers of philosophical concerns. The probability of statistical significance is a function of decisions made by experimenters/analysts. If the decisions are based on convention they are termed arbitrary or mindless while those not so based may be termed subjective. To minimize type II errors, large samples are recommended. In psychology practically all null hypotheses are claimed to be false for sufficiently large samples so "...it is usually nonsensical to perform an experiment with the sole aim of rejecting the null hypothesis.". "Statistically significant findings are often misleading" in psychology. Statistical significance does not imply practical significance and correlation does not imply causation. Casting doubt on the null hypothesis is thus far from directly supporting the research hypothesis.
- "[I]t does not tell us what we want to know". Lists of dozens of complaints are available.

Critics and supporters are largely in factual agreement regarding the characteristics of null hypothesis significance testing (NHST): While it can provide critical information, it is inadequate as the sole tool for statistical analysis. Successfully rejecting the null hypothesis may offer no support for the research hypothesis. The continuing controversy concerns the selection of the best statistical practices for the near-term future given the (often poor) existing practices. Critics would prefer to ban NHST completely, forcing a complete departure from those practices, while supporters suggest a less absolute change.

Controversy over significance testing, and its effects on publication bias in particular, has produced several results. The American Psychological Association has strengthened its statistical reporting requirements after review, medical journal publishers have recognized the obligation to publish some results that are not statistically significant to combat publication bias and a journal (Journal of Articles in Support of the Null Hypothesis) has been created to publish such results exclusively. Textbooks have added some cautions and increased coverage of the tools necessary to estimate the size of the sample required to produce significant results. Major organizations have not abandoned use of significance tests although some have discussed doing so.

Alternatives

The numerous criticisms of significance testing do not lead to a single alternative. A unifying position of critics is that statistics should not lead to a conclusion or a decision but to a probability or to an estimated value with a confidence interval rather than to an accept-reject decision regarding a particular hypothesis. It is unlikely that the controversy surrounding significance testing will be resolved in the near future. Its supposed flaws and unpopularity do not eliminate the need for an objective and transparent means of reaching conclusions regarding studies that produce statistical results. Critics have not unified around an alternative. Other forms of reporting confidence or uncertainty could probably grow in popularity. One strong critic of significance testing suggested a list of reporting alternatives: effect sizes for importance, prediction intervals for confidence, replications and extensions for replicability, meta-analyses for generality. None of these suggested alternatives produces a conclusion/decision. Lehmann said that hypothesis testing theory can be presented in terms of conclusions/decisions, probabilities, or confidence intervals. "The distinction between the ... approaches is largely one of reporting and interpretation."

On one "alternative" there is no disagreement: Fisher himself said, "In relation to the test of significance, we may say that a phenomenon is experimentally demonstrable when we know how to conduct an experiment which will rarely fail to give us a statistically significant result." Cohen, an influential critic of significance testing, concurred, "... don't look for a magic alternative to NHST [null hypothesis significance testing] ... It doesn't exist." "... given the problems of statistical induction, we must finally rely, as have the older sciences, on replication." The "alternative" to significance testing is repeated testing. The easiest way to decrease statistical uncertainty is by obtaining more data, whether by increased sample size or by repeated tests. Nickerson claimed to have never seen the publication of a literally replicated experiment in psychology. An indirect approach to replication is meta-analysis.

Bayesian inference is one proposed alternative to significance testing. (Nickerson cited 10 sources suggesting it, including Rozeboom (1960)).For example, Bayesian parameter estimation can provide rich information about the data from which researchers can draw inferences, while using uncertain priors that exert only minimal influence on the results when enough data is available. Psychologist John K. Kruschke has suggested Bayesian estimation as an alternative for the t-test. Alternatively two competing models/hypothesis can be compared using Bayes factors.Bayesian methods could be criticized for requiring information that is seldom available in the cases where significance testing is most heavily used. Neither the prior probabilities nor the probability distribution of the test statistic under the alternative hypothesis are often available in the social sciences.

Advocates of a Bayesian approach sometimes claim that the goal of a researcher is most often to objectively assess the probability that a hypothesis is true based on the data they have collected. Neither Fisher's significance testing, nor Neyman–Pearson hypothesis testing can provide this information, and do not claim to. The probability a hypothesis is true can only be derived from use of Bayes' Theorem, which was unsatisfactory to both the Fisher and Neyman–Pearson camps due to the explicit use of subjectivity in the form of the prior probability. Fisher's strategy is to sidestep this with the p-value (an objective index based on the data alone) followed by inductive inference, while Neyman–Pearson devised their approach of inductive behaviour.

REFERENCES

1. Bandyopadhyay, P. S.; Forster, M. R., eds. (2011), Philosophy of Statistics, Elsevier.
2. Bickel, Peter J.; Doksum, Kjell A. (2001). Mathematical statistics: Basic and selected topics. 1 (Second (updated printing 2007) ed.). Prentice Hall. ISBN 0-13-850363-X. MR 443141.
3. Cox, D. R. (2006). Principles of Statistical Inference, Cambridge University Press. ISBN 0-521-68567-2.
4. Fisher, R. A. (1955), "Statistical methods and scientific induction", Journal of the Royal Statistical Society, Series B, 17, 69—78. (criticism of statistical theories of Jerzy Neyman and Abraham Wald)
5. Freedman, D. A. (2009). Statistical models: Theory and practice (revised ed.). Cambridge University Press. pp. xiv+442 pp. ISBN 978-0-521-74385-3. MR 2489600.
6. Freedman, D. A. (2010). Statistical Models and Causal Inferences: A Dialogue with the Social Sciences (Edited by David Collier, Jasjeet S. Sekhon, and Philip B. Stark), Cambridge University Press.
7. Hansen, Mark H.; Yu, Bin (June 2001). "Model Selection and the Principle of Minimum Description Length: Review paper". Journal of the American Statistical Association. 96 (454): 746–774. doi:10.1198/016214501753168398. JSTOR 2670311. MR 1939352.
8. Hinkelmann, Klaus; Kempthorne, Oscar (2008). Introduction to Experimental Design (Second ed.). Wiley. ISBN 978-0-471-72756-9.
9. Kolmogorov, Andrei N. (1963). "On tables of random numbers". Sankhyā Ser. A. 25: 369–375. MR 178484. Reprinted as Kolmogorov, Andrei N. (1998). "On tables of random numbers". Theoretical Computer Science. 207 (2): 387–395. doi:10.1016/S0304-3975(98)00075-9. MR 1643414.
10. Konishi S., Kitagawa G. (2008), Information Criteria and Statistical Modeling, Springer.
11. Kruskal, William (December 1988). "Miracles and Statistics: the casual assumption of independence (ASA Presidential Address)". Journal of the American Statistical Association. 83 (404): 929–940. doi:10.2307/2290117. JSTOR 2290117.
12. Le Cam, Lucian. (1986) Asymptotic Methods of Statistical Decision Theory, Springer. ISBN 0-387-96307-3
13. Neyman, Jerzy (1956). "Note on an Article by Sir Ronald Fisher". Journal of the Royal Statistical Society, Series B. 18 (2): 288–294. JSTOR 2983716. (reply to Fisher 1955)
14. Peirce, C. S. (1877–1878), "Illustrations of the Logic of Science" (series), Popular Science Monthly, vols. 12-13. Relevant individual papers:
15. (1878 March), "The Doctrine of Chances", Popular Science Monthly, v. 12, March issue, pp. 604–615. Internet Archive Eprint.

16. (1878 April), "The Probability of Induction", Popular Science Monthly, v. 12, pp. 705–718. Internet Archive Eprint.
17. (1878 June), "The Order of Nature", Popular Science Monthly, v. 13, pp. 203–217.Internet Archive Eprint.
18. (1878 August), "Deduction, Induction, and Hypothesis", Popular Science Monthly, v. 13, pp. 470–482. Internet Archive Eprint.
19. Peirce, C. S. (1883), "A Theory of Probable Inference", Studies in Logic, pp. 126-181, Little, Brown, and Company. (Reprinted 1983, John Benjamins Publishing Company, ISBN 90-272-3271-7)
20. Pfanzagl, Johann; with the assistance of R. Hamböker (1994). Parametric Statistical Theory. Berlin: Walter de Gruyter. ISBN 3-11-013863-8. MR 1291393.
21. Rissanen, Jorma (1989). Stochastic Complexity in Statistical Inquiry. Series in computer science. 15. Singapore: World Scientific. ISBN 9971-5-0859-1. MR 1082556.
22. Soofi, Ehsan S. (December 2000). "Principal Information-Theoretic Approaches (Vignettes for the Year 2000: Theory and Methods, ed. by George Casella)". Journal of the American Statistical Association. 95 (452): 1349–1353. doi:10.1080/01621459.2000.10474346. JSTOR 2669786. MR 1825292.
23. Traub, Joseph F.; Wasilkowski, G. W.; Wozniakowski, H. (1988). Information-Based Complexity. Academic Press. ISBN 0-12-697545-0.
24. Zabell, S. L. (Aug 1992). "R. A. Fisher and Fiducial Argument". Statistical Science. 7 (3): 369–387. doi:10.1214/ss/1177011233. JSTOR 2246073.
25. Hampel, Frank (Feb 2003). "The proper fiducial argument" (PDF) (Research Report No. 114). Retrieved 29 March 2016.
26. Further reading
27. Casella, G., Berger, R.L. (2001). Statistical Inference. Duxbury Press. ISBN 0-534-24312-6
28. Freedman D.A. (1991). "Statistical models and shoe leather", Sociological Methodology, 21: 291–313.
29. Held L., Bové D.S. (2014). Applied Statistical Inference—Likelihood and Bayes (Springer).
30. Lenhard, Johannes (2006). "Models and Statistical Inference: the controversy between Fisher and Neyman–Pearson", British Journal for the Philosophy of Science, 57: 69–91.
31. Lindley, D. (1958). "Fiducial distribution and Bayes' theorem", Journal of the Royal Statistical Society, Series B, 20: 102–7.
32. Rahlf, Thomas (2014). "Statistical Inference", in Claude Diebolt, and Michael Haupert (eds.), "Handbook of Cliometrics (Springer Reference Series)", Berlin/Heidelberg: Springer. http://www.springerreference.com/docs/html/chapterdbid/372458.html
33. Reid, N.; Cox, D. R. (2014). "On Some Principles of Statistical Inference".

International Statistical Review: n/a. doi:10.1111/insr.12067.

34. Young, G.A., Smith, R.L. (2005). Essentials of Statistical Inference, CUP. ISBN 0-521-83971-8
35. Stuart A., Ord K., Arnold S. (1999), Kendall's Advanced Theory of Statistics: Volume 2A—Classical Inference & the Linear Model (Arnold) §20.2.
36. Burnham, K. P.; Anderson, D. R. (2002). Model Selection and Multimodel Inference: A Practical Information-Theoretic Approach (2nd ed.). Springer-Verlag. ISBN 0-387-95364-7..
37. Schervish, M (1996) Theory of Statistics, p. 218. Springer ISBN 0-387-94546-6
38. Kaye, David H.; Freedman, David A. (2011). Reference Guide on Statistics. Reference Manual on Scientific Evidence (3rd ed.). Eagan, MN Washington, D.C: West National Academies Press. p. 259. ISBN 978-0-309-21421-6.
39. Lehmann, E. L.; Romano, Joseph P. (2005). Testing Statistical Hypotheses (3E ed.). New York: Springer. ISBN 0-387-98864-5.
40. Triola, Mario (2001). Elementary statistics (8 ed.). Boston: Addison-Wesley. p. 388. ISBN 0-201-61477-4.
41. Hinkelmann, Klaus and Kempthorne, Oscar (2008). Design and Analysis of Experiments. I and II (Second ed.). Wiley. ISBN 978-0-470-38551-7.
42. Montgomery, Douglas (2009). Design and analysis of experiments. Hoboken, N.J.: Wiley. ISBN 978-0-470-12866-4.
43. R. A. Fisher (1925).Statistical Methods for Research Workers, Edinburgh: Oliver and Boyd, 1925, p.43.
44. Bakan, David (1966). The test of significance in psychological research. Psychological Bulletin. 66 (6): 423–437. doi:10.1037/h0020412.
45. Richard J. Larsen; Donna Fox Stroup (1976). Statistics in the Real World: a book of examples. Macmillan. ISBN 978-0023677205.
46. Hubbard, R.; Parsa, A. R.; Luthy, M. R. (1997). The Spread of Statistical Significance Testing in Psychology: The Case of the Journal of Applied Psychology. Theory and Psychology. 7 (4): 545–554. doi:10.1177/0959354397074006.
47. Moore, David (2003). Introduction to the Practice of Statistics. New York: W.H. Freeman and Co. p. 426. ISBN 9780716796572.
48. Huff, Darrell (1993). How to lie with statistics. New York: Norton. ISBN 0-393-31072-8.
49. Huff, Darrell (1991). How to Lie with Statistics. London: Penguin Books. ISBN 0-14-013629-0.
50. Over the last fifty years, How to Lie with Statistics has sold more copies than any other statistical text. J. M. Steele. Darrell Huff and Fifty Years of How to Lie with Statistics. Statistical Science, 20 (3), 2005, 205–209.
51. Fisher, Sir Ronald A. (1956) [1935]. Mathematics of a Lady Tasting Tea. In James Roy Newman. The World of Mathematics, volume 3 [Design of Experiments]. Courier Dover Publications. ISBN 978-0-486-41151-4. Originally from Fisher›s

book Design of Experiments.

52. Box, Joan Fisher (1978). R.A. Fisher, The Life of a Scientist. New York: Wiley. p. 134. ISBN 0-471-09300-9.
53. C. S. Peirce (August 1878). Illustrations of the Logic of Science VI: Deduction, Induction, and Hypothesis. Popular Science Monthly. 13. Retrieved March 30, 2012.
54. Jaynes, E. T. (2007). Probability theory : the logic of science (5. print. ed.). Cambridge [u.a.]: Cambridge Univ. Press. ISBN 978-0-521-59271-0.
55. Loveland, Jennifer L. (2011). Mathematical Justification of Introductory Hypothesis Tests and Development of Reference Materials (M.Sc. (Mathematics)). Utah State University. Retrieved April 30, 2013. Abstract: The focus was on the Neyman–Pearson approach to hypothesis testing. A brief historical development of the Neyman–Pearson approach is followed by mathematical proofs of each of the hypothesis tests covered in the reference material. The proofs do not reference the concepts introduced by Neyman and Pearson, instead they show that traditional test statistics have the probability distributions ascribed to them, so that significance calculations assuming those distributions are correct. The thesis information is also posted at mathnstats.com as of April 2013.
56. NIST handbook: Two-Sample t-test for Equal Means
57. Steel, R. G. D., and Torrie, J. H., Principles and Procedures of Statistics with Special Reference to the Biological Sciences., McGraw Hill, 1960, page 350.
58. Weiss, Neil A. (1999). Introductory Statistics (5th ed.). p. 802. ISBN 0-201-59877-9.
59. NIST handbook: F-Test for Equality of Two Standard Deviations (Testing standard deviations the same as testing variances)
60. Steel, R. G. D., and Torrie, J. H., Principles and Procedures of Statistics with Special Reference to the Biological Sciences., McGraw Hill, 1960, page 288.)
61. Raymond Hubbard, M. J. Bayarri, P Values are not Error Probabilities. A working paper that explains the difference between Fisher›s evidential p-value and the Neyman–Pearson Type I error rate α.
62. Laplace, P (1778). Memoire Sur Les Probabilities (PDF). Memoirs de l›Academie royale des Sciences de Paris. 9: 227–332.
63. Stigler, Stephen M. (1986). The History of Statistics: The Measurement of Uncertainty before 1900. Cambridge, Mass: Belknap Press of Harvard University Press. p. 134. ISBN 0-674-40340-1.
64. Fisher, R (1955). Statistical Methods and Scientific Induction (PDF). Journal of the Royal Statistical Society, Series B. 17 (1): 69–78.
65. Neyman, J; Pearson, E. S. (January 1, 1933). On the Problem of the most Efficient Tests of Statistical Hypotheses. Philosophical Transactions of the Royal Society A. 231 (694–706): 289–337. doi:10.1098/rsta.1933.0009.
66. Goodman, S N (June 15, 1999). Toward evidence-based medical statistics. 1: The P Value Fallacy. Ann Intern Med. 130 (12): 995–1004. doi:10.7326/0003-

4819-130-12-199906150-00008. PMID 10383371.

67. Lehmann, E. L. (December 1993). The Fisher, Neyman–Pearson Theories of Testing Hypotheses: One Theory or Two?. Journal of the American Statistical Association. 88 (424): 1242–1249. doi:10.1080/01621459.1993.10476404.
68. Fisher, R N (1958). The Nature of Probability (PDF). Centennial Review. 2: 261–274.We are quite in danger of sending highly trained and highly intelligent young men out into the world with tables of erroneous numbers under their arms, and with a dense fog in the place where their brains ought to be. In this century, of course, they will be working on guided missiles and advising the medical profession on the control of disease, and there is no limit to the extent to which they could impede every sort of national effort.
69. Lenhard, Johannes (2006). Models and Statistical Inference: The Controversy between Fisher and Neyman–Pearson. Brit. J. Phil. Sci. 57: 69–91. doi:10.1093/bjps/axi152.
70. Neyman, Jerzy (1967). RA Fisher (1890—1962): An Appreciation.. Science. 156.3781: 1456–1460. doi:10.1126/science.156.3781.1456.
71. Losavich, J. L.; Neyman, J.; Scott, E. L.; Wells, M. A. (1971). Hypothetical explanations of the negative apparent effects of cloud seeding in the Whitetop Experiment.. Proceedings of the U.S. National Academy of Sciences. 68: 2643–2646. doi:10.1073/pnas.68.11.2643.
72. Halpin, P F; Stam, HJ (Winter 2006). Inductive Inference or Inductive Behavior: Fisher and Neyman: Pearson Approaches to Statistical Testing in Psychological Research (1940–1960). The American Journal of Psychology. 119 (4): 625–653. doi:10.2307/20445367. JSTOR 20445367. PMID 17286092.
73. Gigerenzer, Gerd; Zeno Swijtink; Theodore Porter; Lorraine Daston; John Beatty; Lorenz Kruger (1989). Part 3: The Inference Experts. The Empire of Chance: How Probability Changed Science and Everyday Life. Cambridge University Press. pp. 70–122. ISBN 978-0-521-39838-1.
74. Gigerenzer, G (November 2004). Mindless statistics. The Journal of Socio-Economics. 33 (5): 587–606. doi:10.1016/j.socec.2004.09.033.
75. Loftus, G R (1991). On the Tyranny of Hypothesis Testing in the Social Sciences (PDF). Contemporary Psychology. 36 (2): 102–105. doi:10.1037/029395.
76. Meehl, P (1990). Appraising and Amending Theories: The Strategy of Lakatosian Defense and Two Principles That Warrant It (PDF). Psychological Inquiry. 1 (2): 108–141. doi:10.1207/s15327965pli0102_1.
77. Pearson, K (1900). On the criterion that a given system of deviations from the probable in the case of a correlated system of variables is such that it can be reasonably supposed to have arisen from random sampling (PDF). Philosophical Magazine Series. 5 (50): 157–175. doi:10.1080/14786440009463897.
78. Pearson, K (1904). On the Theory of Contingency and Its Relation to Association and Normal Correlation (PDF). Drapers› Company Research Memoirs Biometric Series. 1: 1–35.

79. Zabell, S (1989). R. A. Fisher on the History of Inverse Probability. Statistical Science. 4 (3): 247–256. doi:10.1214/ss/1177012488. JSTOR 2245634.

80. Ash, Robert (1970). Basic probability theory. New York: Wiley. ISBN 978-0471034506.Section 8.2

81. Tukey, John W. (1960). Conclusions vs decisions. Technometrics. 26 (4): 423–433. doi:10.1080/00401706.1960.10489909. Until we go through the accounts of testing hypotheses, separating [Neyman–Pearson] decision elements from [Fisher] conclusion elements, the intimate mixture of disparate elements will be a continual source of confusion. ... There is a place for both doing one›s best and saying only what is certain, but it is important to know, in each instance, both which one is being done, and which one ought to be done.

82. Stigler, Stephen M. (August 1996). The History of Statistics in 1933. Statistical Science. 11 (3): 244–252. doi:10.1214/ss/1032280216. JSTOR 2246117.

83. Berger, James O. (2003). Could Fisher, Jeffreys and Neyman Have Agreed on Testing?. Statistical Science. 18 (1): 1–32. doi:10.1214/ss/1056397485.

84. Morrison, Denton; Henkel, Ramon, eds. (2006) [1970]. The Significance Test Controversy. AldineTransaction. ISBN 0-202-30879-0.

85. Oakes, Michael (1986). Statistical Inference: A Commentary for the Social and Behavioural Sciences. Chichester New York: Wiley. ISBN 0471104434.

86. Chow, Siu L. (1997). Statistical Significance: Rationale, Validity and Utility. ISBN 0-7619-5205-5.

87. Harlow, Lisa Lavoie; Stanley A. Mulaik; James H. Steiger, eds. (1997). What If There Were No Significance Tests?. Lawrence Erlbaum Associates. ISBN 978-0-8058-2634-0.

88. Kline, Rex (2004). Beyond Significance Testing: Reforming Data Analysis Methods in Behavioral Research. Washington, D.C.: American Psychological Association. ISBN 9781591471189.

89. McCloskey, Deirdre N.; Stephen T. Ziliak (2008). The Cult of Statistical Significance: How the Standard Error Costs Us Jobs, Justice, and Lives. University of Michigan Press. ISBN 0-472-05007-9.

90. Cornfield, Jerome (1976). Recent Methodological Contributions to Clinical Trials (PDF). American Journal of Epidemiology. 104 (4): 408–421.

91. Yates, Frank (1951). The Influence of Statistical Methods for Research Workers on the Development of the Science of Statistics. Journal of the American Statistical Association. 46: 19–34. doi:10.1080/01621459.1951.10500764. The emphasis given to formal tests of significance throughout [R.A. Fisher›s] Statistical Methods ... has caused scientific research workers to pay undue attention to the results of the tests of significance they perform on their data, particularly data derived from experiments, and too little to the estimates of the magnitude of the effects they are investigating. ... The emphasis on tests of significance and the consideration of the results of each experiment in isolation, have had the unfortunate consequence that scientific workers have often regarded the execution of a test of significance on an experiment as the ultimate objective.

92. Begg, Colin B.; Berlin, Jesse A. (1988). Publication bias: a problem in interpreting medical data. Journal of the Royal Statistical Society, Series A: 419–463.

93. Meehl, Paul E. (1967). Theory-Testing in Psychology and Physics: A Methodological Paradox (PDF). Philosophy of Science. 34 (2): 103–115. doi:10.1086/288135. Thirty years later, Meehl acknowledged statistical significance theory to be mathematically sound while continuing to question the default choice of null hypothesis, blaming instead the social scientists› poor understanding of the logical relation between theory and fact in The Problem Is Epistemology, Not Statistics: Replace Significance Tests by Confidence Intervals and Quantify Accuracy of Risky Numerical Predictions (Chapter 14 in Harlow (1997)).

94. Nunnally, Jum (1960). The place of statistics in psychology. Educational and Psychological Measurement. 20 (4): 641–650. doi:10.1177/001316446002000401.

95. Lykken, David T. (1991). What›s wrong with psychology, anyway?. Thinking Clearly About Psychology. 1: 3–39.

96. Jacob Cohen (December 1994). The Earth Is Round ($p < .05$). American Psychologist. 49 (12): 997–1003. doi:10.1037/0003-066X.49.12.997. This paper lead to the review of statistical practices by the APA. Cohen was a member of the Task Force that did the review.

97. Nickerson, Raymond S. (2000). Null Hypothesis Significance Tests: A Review of an Old and Continuing Controversy. Psychological Methods. 5 (2): 241–301. doi:10.1037/1082-989X.5.2.241. PMID 10937333.

98. Branch, Mark (2014). Malignant side effects of null hypothesis significance testing. Theory & Psychology. 24 (2): 256–277. doi:10.1177/0959354314525282.

99. Wilkinson, Leland (1999). Statistical Methods in Psychology Journals; Guidelines and Explanations. American Psychologist. 54 (8): 594–604. doi:10.1037/0003-066X.54.8.594. Hypothesis tests. It is hard to imagine a situation in which a dichotomous accept-reject decision is better than reporting an actual p value or, better still, a confidence interval. (p 599). The committee used the cautionary term forbearance in describing its decision against a ban of hypothesis testing in psychology reporting. (p 603)

100. ICMJE: Obligation to Publish Negative Studies. Retrieved September 3, 2012. Editors should seriously consider for publication any carefully done study of an important question, relevant to their readers, whether the results for the primary or any additional outcome are statistically significant. Failure to submit or publish findings because of lack of statistical significance is an important cause of publication bias.

101. Journal of Articles in Support of the Null Hypothesis website: JASNH homepage. Volume 1 number 1 was published in 2002, and all articles are on psychology-related subjects.

102. Howell, David (2002). Statistical Methods for Psychology (5 ed.). Duxbury. p. 94. ISBN 0-534-37770-X.

103. Armstrong, J. Scott (2007). Significance tests harm progress in forecasting. International Journal of Forecasting. 23 (2): 321–327. doi:10.1016/j.

ijforecast.2007.03.004.

104. E. L. Lehmann (1997). Testing Statistical Hypotheses: The Story of a Book. Statistical Science. 12 (1): 48–52. doi:10.1214/ss/1029963261.
105. Kruschke, J K (July 9, 2012). Bayesian Estimation Supersedes the T Test. Journal of Experimental Psychology: General. N/A (N/A): N/A. doi:10.1037/a0029146.
106. Kass, R. E. (1993). Bayes factors and model uncertainty (PDF). Department of Statistics, University of Washington.
107. Rozeboom, William W (1960). The fallacy of the null-hypothesis significance test (PDF). Psychological Bulletin. 57 (5): 416–428. doi:10.1037/h0042040. ...the proper application of statistics to scientific inference is irrevocably committed to extensive consideration of inverse [AKA Bayesian] probabilities... It was acknowledged, with regret, that a priori probability distributions were available only as a subjective feel, differing from one person to the next in the more immediate future, at least.
108. Berger, James (2006). The Case for Objective Bayesian Analysis. Bayesian Analysis. 1 (3): 385–402. doi:10.1214/06-ba115. In listing the competing definitions of objective Bayesian analysis, A major goal of statistics (indeed science) is to find a completely coherent objective Bayesian methodology for learning from data. The author expressed the view that this goal is not attainable.
109. Aldrich, J (2008). R. A. Fisher on Bayes and Bayes› theorem (PDF). Bayesian Analysis. 3 (1): 161–170. doi:10.1214/08-BA306.
110. Mayo, D. G.; Spanos, A. (2006). Severe Testing as a Basic Concept in a Neyman–Pearson Philosophy of Induction. The British Journal for the Philosophy of Science. 57 (2): 323. doi:10.1093/bjps/axl003.
111. Mathematics > High School: Statistics & Probability > Introduction Common Core State Standards Initiative (relates to USA students)
112. College Board Tests > AP: Subjects > Statistics The College Board (relates to USA students)
113. Huff, Darrell (1993). How to lie with statistics. New York: Norton. p. 8. ISBN 0-393-31072-8.›Statistical methods and statistical terms are necessary in reporting the mass data of social and economic trends, business conditions, opinion polls, the census. But without writers who use the words with honesty and readers who know what they mean, the result can only be semantic nonsense.›
114. Snedecor, George W.; Cochran, William G. (1967). Statistical Methods (6 ed.). Ames, Iowa: Iowa State University Press. p. 3. ...the basic ideas in statistics assist us in thinking clearly about the problem, provide some guidance about the conditions that must be satisfied if sound inferences are to be made, and enable us to detect many inferences that have no good logical foundation.
115. Sotos, Ana Elisa Castro; Vanhoof, Stijn; Noortgate, Wim Van den; Onghena, Patrick (2007). Students› Misconceptions of Statistical Inference: A Review of the Empirical Evidence from Research on Statistics Education. Educational Research Review. 2: 98–113. doi:10.1016/j.edurev.2007.04.001.

116. Moore, David S. (1997). New Pedagogy and New Content: The Case of Statistics. International Statistical Review. 65: 123–165. doi:10.2307/1403333.

117. Hubbard, Raymond; Armstrong, J. Scott (2006). Why We Don›t Really Know What Statistical Significance Means: Implications for Educators (PDF). Journal of Marketing Education. 28 (2): 114. doi:10.1177/0273475306288399. Archived from the original on May 18, 2006. Preprint

118. Sotos, Ana Elisa Castro; Vanhoof, Stijn; Noortgate, Wim Van den; Onghena, Patrick (2009). How Confident Are Students in Their Misconceptions about Hypothesis Tests?. Journal of Statistics Education. 17 (2).

119. Gigerenzer, G. (2004). The Null Ritual What You Always Wanted to Know About Significant Testing but Were Afraid to Ask (PDF). The SAGE Handbook of Quantitative Methodology for the Social Sciences. pp. 391–408. doi:10.4135/9781412986311.

CHAPTER 5

Correlation and Dependence

Chapter Contents

- **Introduction**
- **Pearson's Product-Moment Coefficient**
- **Pearson Product-Moment Correlation Coefficient**
- **Confounding**
- **Coefficient Of Determination**

Correlation is a statistical measure that indicates the extent to which two or more variables fluctuate together. A positive correlation indicates the extent to which those variables increase or decrease in parallel; a negative correlation indicates the extent to which one variable increases as the other decreases.

A correlation coefficient is a statistical measure of the degree to which changes to the value of one variable predict change to the value of another. When the fluctuation of one variable reliably predicts a similar fluctuation in another variable, there's often a tendency to think that means that the change in one causes the change in the other. However, correlation does not imply causation. There may be, for example, an unknown factor that influences both variables similarly.

INTRODUCTION

In statistics, dependence or association is any statistical relationship, whether causal or not, between two random variables or two sets of data. Correlation is any of a broad class of statistical relationships involving dependence, though in common usage it most often refers to the extent to which two variables have a linear relationship with each other. Familiar examples of dependent phenomena include the correlation between the physical statures of parents and their offspring, and the correlation between the demand for a product and its price.

Correlations are useful because they can indicate a predictive relationship that can be exploited in practice. For example, an electrical utility may produce less power on a mild day based on the correlation between electricity demand and weather. In this example there is a causal relationship, because extreme weather causes people to use more electricity for heating or cooling; however, correlation is not sufficient to demonstrate the presence of such a causal relationship (i.e., correlation does not imply causation).

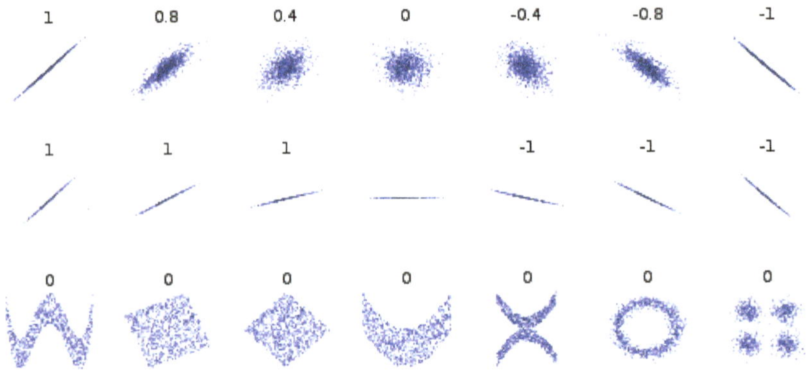

Several sets of (x, y) points, with the Pearson correlation coefficient of x and y for each set. Note that the correlation reflects the noisiness and direction of a linear relationship (top row), but not the slope of that relationship (middle), nor many aspects of nonlinear relationships (bottom). N.B.: the figure in the center has a slope of 0 but in that case the correlation coefficient is undefined because the variance of Y is zero.

Formally, dependence refers to any situation in which random variables do not satisfy a mathematical condition of probabilistic independence. In loose usage, correlation can refer to any departure of two or more random variables from independence, but technically it refers to any of several more specialized types of relationship between mean values. There are several correlation coefficients, often denoted ρ or r, measuring the degree of correlation. The most common of these is the Pearson correlation coefficient, which is sensitive only to a linear relationship between two variables (which may exist even if one is a nonlinear function of the other). Other correlation coefficients have been developed to be more robust than the Pearson correlation – that is, more sensitive to nonlinear relationships. Mutual information can also be applied to measure dependence between two variables.

PEARSON'S PRODUCT-MOMENT COEFFICIENT

The most familiar measure of dependence between two quantities is the Pearson product-moment correlation coefficient, or "Pearson's correlation coefficient", commonly called simply "the correlation coefficient". It is obtained by dividing the covariance of the two variables by the product of their standard deviations. Karl Pearson developed the coefficient from a similar but slightly different idea by Francis Galton.

The population correlation coefficient $\rho_{X,Y}$ between two random variables X and Y with expected values μ_X and μ_Y and standard deviations σ_X and σ_Y is defined as:

$$\rho_{X,Y} = \mathrm{corr}(X,Y) = \frac{\mathrm{cov}(X,Y)}{\sigma_X \sigma_Y} = \frac{E[(X-\mu_X)(Y-\mu_Y)]}{\sigma_X \sigma_Y},$$

where E is the expected value operator, cov means covariance, and corr is a widely used alternative notation for the correlation coefficient.

The Pearson correlation is defined only if both of the standard deviations are finite and nonzero. It is a corollary of the Cauchy–Schwarz inequality that the correlation cannot exceed 1 in absolute value. The correlation coefficient is symmetric: corr(X,Y) = corr(Y,X).

The Pearson correlation is +1 in the case of a perfect direct (increasing) linear relationship (correlation), −1 in the case of a perfect decreasing (inverse) linear relationship (anticorrelation), and some value in the open interval (−1, 1) in all other cases, indicating the degree of linear dependence between the variables. As it approaches zero there is less of a relationship (closer to uncorrelated). The closer the coefficient is to either −1 or 1, the stronger the correlation between the variables.

If the variables are independent, Pearson's correlation coefficient is 0, but the converse is not true because the correlation coefficient detects only linear dependencies between two variables. For example, suppose the random variable X is symmetrically distributed about zero, and $Y = X^2$. Then Y is completely determined by X, so that X and Y are perfectly dependent, but their correlation is zero; they are uncorrelated. However, in the special case when X and Y are jointly normal, uncorrelatedness is equivalent to independence.

If we have a series of n measurements of X and Y written as x_i and y_i for i = 1, 2, ..., n, then the sample correlation coefficient can be used to estimate the population Pearson correlation r between X and Y. The sample correlation coefficient is written:

$$r_{xy} = \frac{\sum_{i=1}^{n}(x_i - \bar{x})(y_i - \bar{y})}{(n-1)s_x s_y} = \frac{\sum_{i=1}^{n}(x_i - \bar{x})(y_i - \bar{y})}{\sqrt{\sum_{i=1}^{n}(x_i - \bar{x})^2 \sum_{i=1}^{n}(y_i - \bar{y})^2}},$$

where x and y are the sample means of X and Y, and s_x and s_y are the sample standard deviations of X and Y.

This can also be written as:

$$r_{xy} = \frac{\sum x_i y_i - n\bar{x}\bar{y}}{n s_x s_y}$$

$$= \frac{n\sum x_i y_i - \sum x_i \sum y_i}{\sqrt{n\sum x_i^2 - (\sum x_i)^2}\sqrt{n\sum y_i^2 - (\sum y_i)^2}}.$$

If x and y are results of measurements that contain measurement error, the realistic limits on the correlation coefficient are not −1 to +1 but a smaller range.

For the case of a linear model with a single independent variable, the coefficient of determination (R squared) is the square of r, Pearson's product-moment coefficient.

Rank Correlation Coefficients

Rank correlation coefficients, such as Spearman's rank correlation coefficient and Kendall's rank correlation coefficient (τ) measure the extent to which, as one variable increases, the other variable tends to increase, without requiring that increase to be represented by a linear relationship. If, as the one variable increases, the other decreases, the rank correlation coefficients will be negative. It is common to regard these rank correlation coefficients as alternatives to Pearson's coefficient, used either to reduce the amount of calculation or to make the coefficient less sensitive to non-normality in distributions. However, this view has little mathematical basis, as rank correlation coefficients measure a different type of relationship than the Pearson product-moment correlation coefficient, and are best seen as measures of a different type of association, rather than as alternative measure of the population correlation coefficient.

To illustrate the nature of rank correlation, and its difference from linear correlation, consider the following four pairs of numbers (x, y):

(0, 1), (10, 100), (101, 500), (102, 2000).

As we go from each pair to the next pair x increases, and so does y. This relationship is perfect, in the sense that an increase in x is always accompanied by an increase in y. This means that we have a perfect rank correlation, and both Spearman's and Kendall's correlation coefficients are 1, whereas in this example Pearson product-moment correlation coefficient is 0.7544, indicating that the points are far from lying on a straight line. In the same way if y always decreases when x increases, the rank correlation coefficients will be −1, while the Pearson product-moment correlation coefficient may or may not be close to −1, depending on how close the points are to a straight line. Although in the extreme cases of perfect rank correlation the two coefficients are both equal (being both +1 or both −1), this is not generally the case, and so values of the two coefficients cannot meaningfully be compared. For example, for the three pairs (1, 1) (2, 3) (3, 2) Spearman's coefficient is 1/2, while Kendall's coefficient is 1/3.

Other Measures of Dependence among Random Variables

The information given by a correlation coefficient is not enough to define the dependence structure between random variables. The correlation coefficient completely defines the dependence structure only in very particular cases, for example when the distribution is a multivariate normal distribution. (See diagram above.) In the case of elliptical distributions it characterizes the (hyper-)ellipses of equal density, however, it does not completely characterize the dependence structure (for example, a multivariate t-distribution's degrees of freedom determine the level of tail dependence). Distance correlation was introduced to address the deficiency of Pearson's correlation that it can be zero for dependent random variables; zero distance correlation implies independence. The Randomized Dependence Coefficient is a computationally efficient, copula-based measure of dependence between multivariate random variables. RDC is invariant with respect to non-linear scalings of random variables, is capable of discovering a wide range of functional association patterns and takes value zero at independence.

The correlation ratio is able to detect almost any functional dependency, and the entropy-based mutual information, total correlation and dual total correlation are capable of detecting even more general dependencies. These are sometimes referred to as multi-moment correlation measures, in comparison to those that consider only second moment (pairwise or quadratic) dependence.

The polychoric correlation is another correlation applied to ordinal data that aims to estimate the correlation between theorised latent variables.

One way to capture a more complete view of dependence structure is to consider a copula between them.

The coefficient of determination generalizes the correlation coefficient for relationships beyond simple linear regression.

Sensitivity to the Data Distribution

The degree of dependence between variables X and Y does not depend on the scale on which the variables are expressed. That is, if we are analyzing the relationship between X and Y, most correlation measures are unaffected by transforming X to a + bX and Y to c + dY, where a, b, c, and d are constants (b and d being positive). This is true of some correlation statistics as well as their population analogues. Some correlation statistics, such as the rank correlation coefficient, are also invariant to monotone transformations of the marginal distributions of X and/or Y.

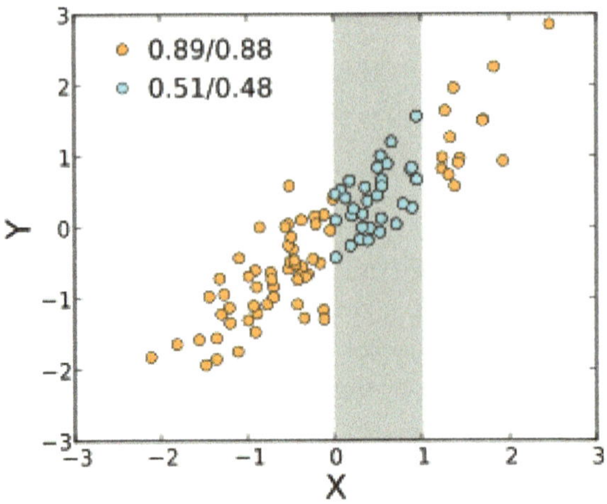

Pearson/Spearman correlation coefficients between X and Y are shown when the two variables' ranges are unrestricted, and when the range of X is restricted to the interval (0,1).

Most correlation measures are sensitive to the manner in which X and Y are sampled. Dependencies tend to be stronger if viewed over a wider range of values. Thus, if we consider the correlation coefficient between the heights of fathers and their sons over all adult males, and compare it to the same correlation coefficient calculated when the fathers are selected to be between 165 cm and 170 cm in height, the correlation will be weaker in the latter case. Several techniques have been developed that attempt to correct for range restriction in one or both variables, and are commonly used in meta-analysis; the most common are Thorndike's case II and case III equations.

Various correlation measures in use may be undefined for certain joint distributions of X and Y. For example, the Pearson correlation coefficient is defined in terms of moments, and hence will be undefined if the moments are undefined. Measures of dependence based on quantiles are always defined. Sample-based statistics intended to estimate population measures of dependence may or may not have desirable statistical properties such as being unbiased, or asymptotically consistent, based on the spatial structure of the population from which the data were sampled.

Sensitivity to the data distribution can be used to an advantage. For example, scaled correlation is designed to use the sensitivity to the range in order to pick out correlations between fast components of time series. By reducing the range of values in a controlled manner, the correlations on long time scale are filtered out and only the correlations on short time scales are revealed.

Correlation Matrices

The correlation matrix of n random variables $X_1, ..., X_n$ is the n × n matrix whose i,j entry is corr(X_i, X_j). If the measures of correlation used are product-moment coefficients, the correlation matrix is the same as the covariance matrix of the standardized random variables $X_i / \sigma (X_i)$ for i = 1, ..., n. This applies to both the matrix of population correlations (in which case "σ" is the population standard deviation), and to the matrix of sample correlations (in which case "σ" denotes the sample standard deviation). Consequently, each is necessarily a positive-semidefinite matrix.

The correlation matrix is symmetric because the correlation between X_i and X_j is the same as the correlation between X_j and X_i.

Common Misconceptions

Correlation and causality

The conventional dictum that "correlation does not imply causation" means that correlation cannot be used to infer a causal relationship between the variables.This dictum should not be taken to mean that correlations cannot indicate the potential existence of causal relations. However, the causes underlying the correlation, if any, may be indirect and unknown, and high correlations also overlap with identity relations (tautologies), where no causal process exists. Consequently, establishing a correlation between two variables is not a sufficient condition to establish a causal relationship (in either direction).

A correlation between age and height in children is fairly causally transparent, but a correlation between mood and health in people is less so. Does improved mood lead to improved health, or does good health lead to good mood, or both? Or does some other factor underlie both? In other words, a correlation can be taken as evidence for a possible causal relationship, but cannot indicate what the causal relationship, if any, might be.

Correlation and Linearity

The Pearson correlation coefficient indicates the strength of a linear relationship between two variables, but its value generally does not completely characterize their relationship. In particular, if the conditional mean of Y given X, denoted E(Y | X), is not linear in X, the correlation coefficient will not fully determine the form of E(Y | X).

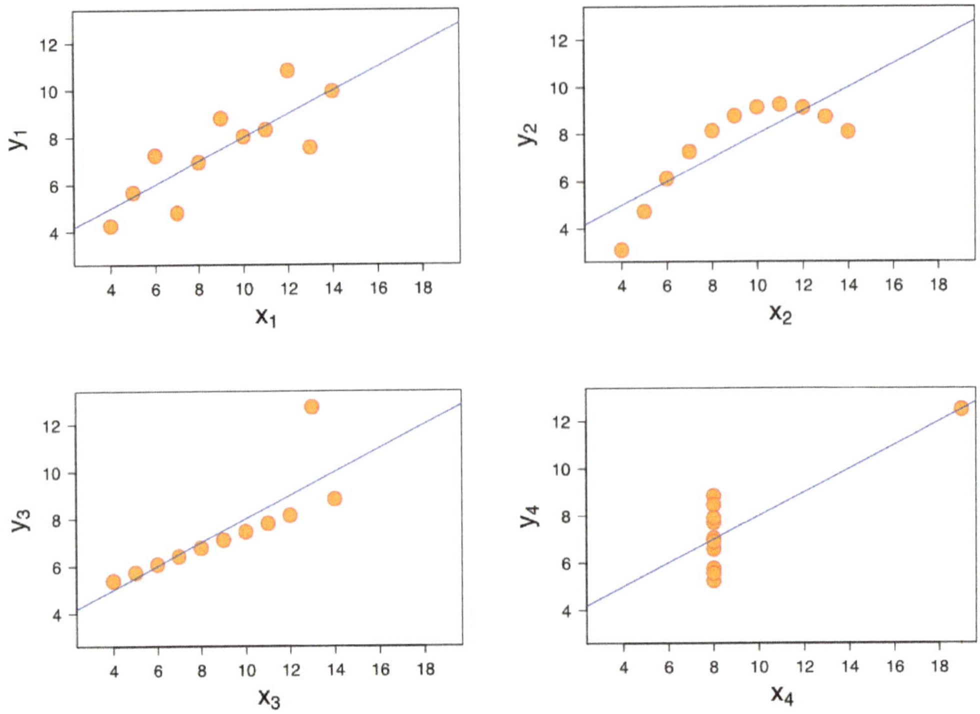

Four sets of data with the same correlation of 0.816

The image on the right shows scatter plots of Anscombe's quartet, a set of four different pairs of variables created by Francis Anscombe. The four y variables have the same mean (7.5), variance (4.12), correlation (0.816) and regression line (y = 3 + 0.5x). However, as can be seen on the plots, the distribution of the variables is very different. The first one (top left) seems to be distributed normally, and corresponds to what one would expect when considering two variables correlated and following the assumption of normality. The second one (top right) is not distributed normally; while an obvious relationship between the two variables can be observed, it is not linear. In this case the Pearson correlation coefficient does not indicate that there is an exact functional relationship: only the extent to which that relationship can be approximated by a linear relationship. In the third case (bottom left), the linear relationship is perfect, except for one outlier which exerts enough influence to lower the correlation coefficient from 1 to 0.816. Finally, the fourth example (bottom right) shows another example when one outlier is enough to produce a high correlation coefficient, even though the relationship between the two variables is not linear.

These examples indicate that the correlation coefficient, as a summary statistic, cannot replace visual examination of the data. Note that the examples are sometimes said to demonstrate that the Pearson correlation assumes that the data follow a normal distribution, but this is not correct.

Bivariate Normal Distribution

If a pair (X, Y) of random variables follows a bivariate normal distribution, the conditional mean E(X|Y) is a linear function of Y, and the conditional mean E(Y|X) is a linear function of X. The correlation coefficient r between X and Y, along with the marginal means and variances of X and Y, determines this linear relationship:

$$E(Y \mid X) = E(Y) + r\sigma_y \frac{X - E(X)}{\sigma_x},$$

where $E(X)$ and $E(Y)$ are the expected values of X and Y, respectively, and σ_x and σ_y are the standard deviations of X and Y, respectively.

Partial Correlation

If a population or data-set is characterized by more than two variables, a partial correlation coefficient measures the strength of dependence between a pair of variables that is not accounted for by the way in which they both change in response to variations in a selected subset of the other variables.

PEARSON PRODUCT-MOMENT CORRELATION COEFFICIENT

In statistics, the Pearson product-moment correlation coefficient (sometimes referred to as the PPMCC or PCC or Pearson's r) is a measure of the linear dependence between two variables X and Y. It has a value between +1 and −1 inclusive, where 1 is total positive linear correlation, 0 is no linear correlation, and −1 is total negative linear correlation. It is widely used in the sciences. It was developed by Karl Pearson from a related idea introduced by Francis Galton in the 1880s. Early work on the distribution of the sample correlation coefficient was carried out by Anil Kumar Gain and R. A. Fisher from the University of Cambridge.

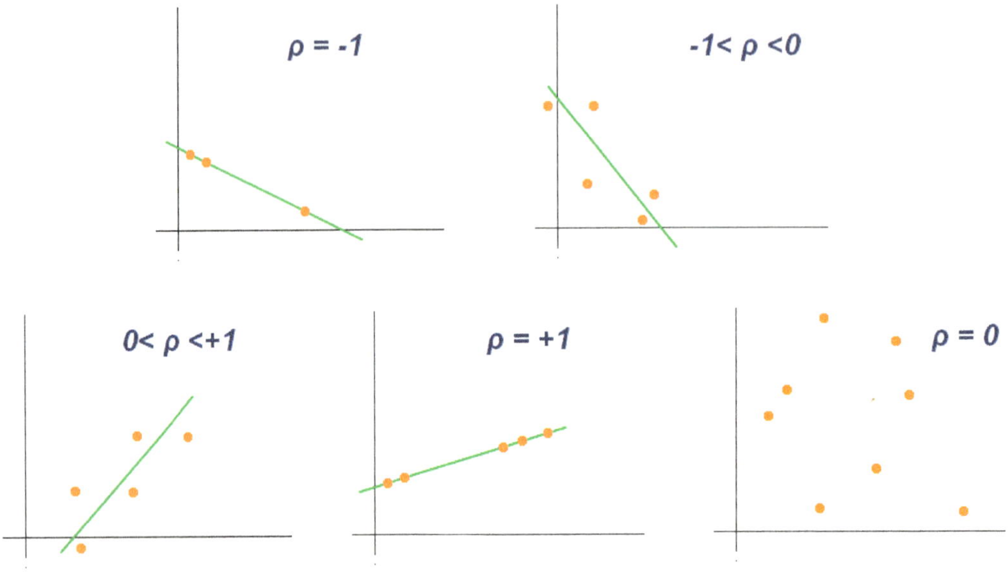

Examples of scatter diagrams with different values of correlation coefficient (ρ)

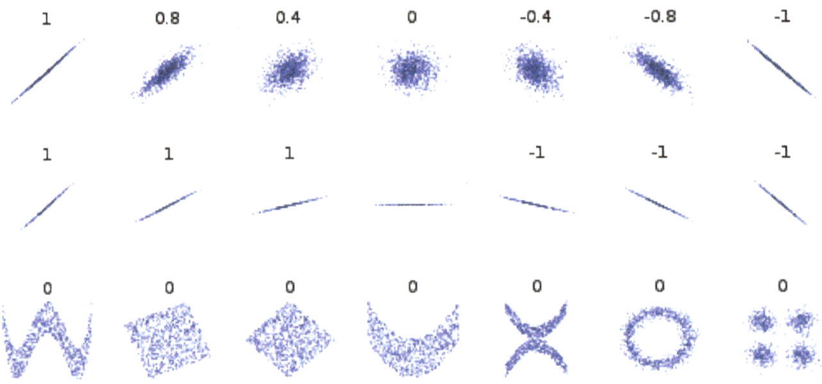

Several sets of (x, y) points, with the correlation coefficient of x and y for each set. Note that the correlation reflects the non-linearity and direction of a linear relationship (top row), but not the slope of that relationship (middle), nor many aspects of nonlinear relationships (bottom). N.B.: the figure in the center has a slope of 0 but in that case the correlation coefficient is undefined because the variance of Y is zero.

Definition

Pearson's correlation coefficient is the covariance of the two variables divided by the product of their standard deviations. The form of the definition involves a "product moment", that is, the mean (the first moment about the origin) of the product of the mean-adjusted random variables; hence the modifier product-moment in the name.

For a Population

Pearson's correlation coefficient when applied to a population is commonly represented by the Greek letter ρ (rho) and may be referred to as the population correlation coefficient or the population Pearson correlation coefficient. The formula for ρ is:

$$\rho_{X,Y} = \frac{\mathrm{cov}(X,Y)}{\sigma_X \sigma_Y}$$

where:

- cov is the covariance
- σ_X is the standard deviation of X

The formula for ρ can be expressed in terms of mean and expectation. Since

- $\mathrm{cov}(X,Y) = \mathrm{E}[(X-\mu_X)(Y-\mu_Y)]$

Then the formula for ρ can also be written as

$$\rho_{X,Y} = \frac{\mathrm{E}[(X-\mu_X)(Y-\mu_Y)]}{\sigma_X \sigma_Y}$$

where:

- cov and σ_X are defined as above
- μ_X is the mean of X
- E is the expectation.

The formula for ρ can be expressed in terms of uncentered moments. Since

- $\mu_X = \mathrm{E}[X]$
- $\mu_Y = \mathrm{E}[Y]$
- $\sigma_X^2 = \mathrm{E}[(X-\mathrm{E}[X])^2] = \mathrm{E}[X^2] - [E[X]]^2$
- $\sigma_Y^2 = \mathrm{E}[(Y-\mathrm{E}[Y])^2] = \mathrm{E}[Y^2] - [E[Y]]^2$
- $\mathrm{E}[(X-\mu_X)(Y-\mu_Y)] = \mathrm{E}[(X-\mathrm{E}[X])(Y-\mathrm{E}[Y])] = \mathrm{E}[XY] - \mathrm{E}[X]\,\mathrm{E}[Y],$

the formula for ρ can also be written as

$$\rho_{X,Y} = \frac{\mathrm{E}[XY] - \mathrm{E}[X]\,\mathrm{E}[Y]}{\sqrt{\mathrm{E}[X^2] - [E[X]]^2}\,\sqrt{\mathrm{E}[Y^2] - [E[Y]]^2}}.$$

For a Sample

Pearson's correlation coefficient when applied to a sample is commonly represented by the letter r and may be referred to as the sample correlation coefficient or the sample

Pearson correlation coefficient. We can obtain a formula for r by substituting estimates of the covariances and variances based on a sample into the formula above. So if we have one dataset $\{x_1,...,x_n\}$ containing n values and another dataset $\{y_1,...,y_n\}$ containing n values then that formula for r is:

$$r = \frac{\sum_{i=1}^{n}(x_i - \bar{x})(y_i - \bar{y})}{\sqrt{\sum_{i=1}^{n}(x_i - \bar{x})^2}\sqrt{\sum_{i=1}^{n}(y_i - \bar{y})^2}}$$

where:

- n, x_i, y_i are defined as above
- $\bar{x} = \frac{1}{n}\sum_{i=1}^{n} x_i$ (the sample mean); and analogously for $\bar{y}$

Rearranging gives us this formula for r:

$$r = r_{xy} = \frac{n\sum x_i y_i - \sum x_i \sum y_i}{\sqrt{n\sum x_i^2 - (\sum x_i)^2}\ \sqrt{n\sum y_i^2 - (\sum y_i)^2}}.$$

where:

- n, x_i, y_i are defined as above
- This formula suggests a convenient single-pass algorithm for calculating sample correlations, but, depending on the numbers involved, it can sometimes be numerically unstable.

Rearranging again gives us this formula for r:

$$r = r_{xy} = \frac{\sum x_i y_i - n\bar{x}\bar{y}}{\sqrt{(\sum x_i^2 - n\bar{x}^2)}\ \sqrt{(\sum y_i^2 - n\bar{y}^2)}}.$$

where:

- $n, x_i, y_i, \bar{x}, \bar{y}$ are defined as above

An equivalent expression gives the formula for r as the mean of the products of the standard scores as follows:

$$r = r_{xy} = \frac{1}{n-1}\sum_{i=1}^{n}\left(\frac{x_i - \bar{x}}{s_x}\right)\left(\frac{y_i - \bar{y}}{s_y}\right)$$

where

- $n, x_i, y_i, \bar{x}, \bar{y}$ are defined as above, and s_x, s_y are defined below
- $\left(\frac{x_i - \bar{x}}{s_x}\right)$ is the standard score (and analogously for the standard score of y)

Alternative formulae for r are also available. One can use the following formula for r:

$$r = r_{xy} = \frac{\sum x_i y_i - n\bar{x}\bar{y}}{(n-1)s_x s_y}$$

where:

- $n, x_i, y_i, \bar{x}, \bar{y}$ are defined as above and:
- $s_x = \sqrt{\frac{1}{n-1}\sum_{i=1}^{n}(x_i - \bar{x})^2}$ (the sample standard deviation); and analogously for s_y

Mathematical Properties

The absolute values of both the sample and population Pearson correlation coefficients are less than or equal to 1. Correlations equal to 1 or −1 correspond to data points lying exactly on a line (in the case of the sample correlation), or to a bivariate distribution entirely supported on a line (in the case of the population correlation). The Pearson correlation coefficient is symmetric: corr(X,Y) = corr(Y,X).

A key mathematical property of the Pearson correlation coefficient is that it is invariant to separate changes in location and scale in the two variables. That is, we may transform X to a + bX and transform Y to c + dY, where a, b, c, and d are constants with b, d ≠ 0, without changing the correlation coefficient. (This fact holds for both the population and sample Pearson correlation coefficients.)

Interpretation

The correlation coefficient ranges from −1 to 1. A value of 1 implies that a linear equation describes the relationship between X and Y perfectly, with all data points lying on a line for which Y increases as X increases. A value of −1 implies that all data points lie on a line for which Y decreases as X increases. A value of 0 implies that there is no linear correlation between the variables.

More generally, note that $(X_i - X)(Y_i - Y)$ is positive if and only if X_i and Y_i lie on the same side of their respective means. Thus the correlation coefficient is positive if X_i and Y_i tend to be simultaneously greater than, or simultaneously less than, their respective means. The correlation coefficient is negative if X_i and Y_i tend to lie on opposite sides of their respective means. Moreover, the stronger is either tendency, the larger is the absolute value of the correlation coefficient.

Geometric Interpretation

For uncentered data, it is possible to obtain a relation between correlation coefficient and the angle φ between both the two regression lines, $y=g_x(x)$ and $x=g_y(y)$, obtained by regressing y on x, and x on y, respectively. One can show that $r = \sec\varphi - \tan\varphi$.

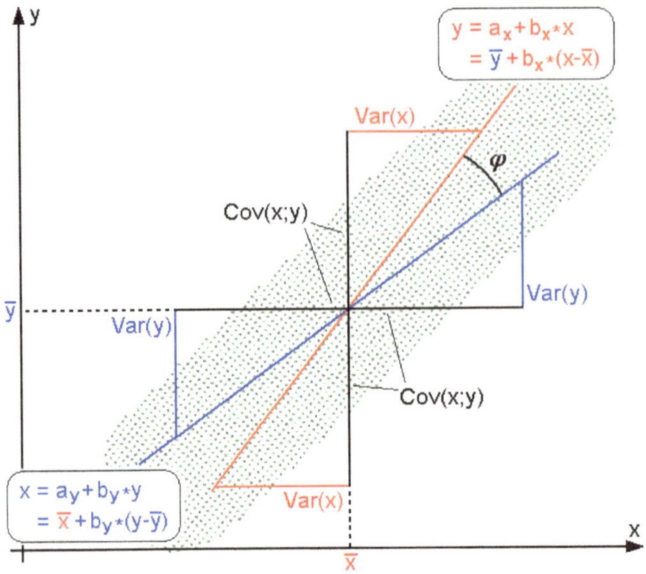

Regression lines for y=g_x(x) [red] and x=g_y(y) [blue]

For centered data (i.e., data which have been shifted by the sample means of their respective variables so as to have an average of zero for each variable), the correlation coefficient can also be viewed as the cosine of the angle q between the two vectors of samples in N-dimensional space (for N samples of each variable) (as illustrated for a special case in the next paragraph).

Both the uncentered (non-Pearson-compliant) and centered correlation coefficients can be determined for a dataset. As an example, suppose five countries are found to have gross national products of 1, 2, 3, 5, and 8 billion dollars, respectively. Suppose these same five countries (in the same order) are found to have 11%, 12%, 13%, 15%, and 18% poverty. Then let **x** and **y** be ordered 5-element vectors containing the above data: **x** = (1, 2, 3, 5, 8) and **y** = (0.11, 0.12, 0.13, 0.15, 0.18).

By the usual procedure for finding the angle q between two vectors (see dot product), the uncentered correlation coefficient is:

$$\cos\theta = \frac{\mathbf{x}\cdot\mathbf{y}}{\|\mathbf{x}\|\,\|\mathbf{y}\|} = \frac{2.93}{\sqrt{103}\sqrt{0.0983}} = 0.920814711.$$

Note that the above data were deliberately chosen to be perfectly correlated: y = 0.10 + 0.01 x. The Pearson correlation coefficient must therefore be exactly one. Centering the data (shifting **x** by E(**x**) = 3.8 and **y** by E(**y**) = 0.138) yields **x** = (−2.8, −1.8, −0.8, 1.2, 4.2) and **y** = (−0.028, −0.018, −0.008, 0.012, 0.042), from which

$$\cos\theta = \frac{\mathbf{x}\cdot\mathbf{y}}{\|\mathbf{x}\|\,\|\mathbf{y}\|} = \frac{0.308}{\sqrt{30.8}\sqrt{0.00308}} = 1 = \rho_{xy},$$

as expected.

Interpretation of the Size of a Correlation

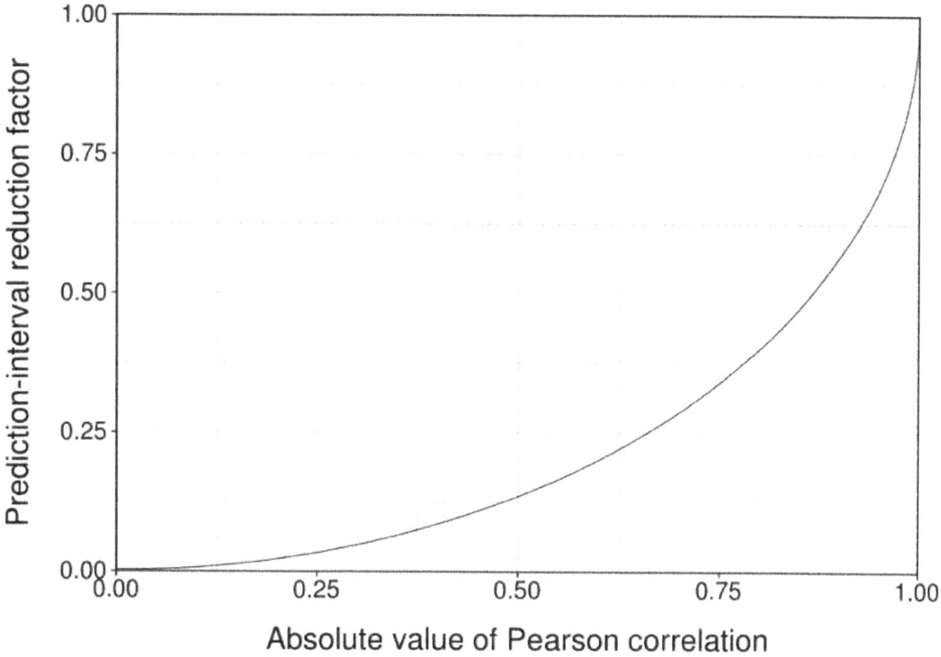

This figure gives a sense of how the usefulness of a Pearson correlation for predicting values varies with its magnitude. Given jointly normal X, Y with correlation ρ, $1-\sqrt{1-\rho^2}$ (plotted here as a function of ρ) is the factor by which a given prediction interval for Y may be reduced given the corresponding value of X. For example, if ρ = .5, then the 95% prediction interval of Y|X will be about 13% smaller than the 95% prediction interval of Y.

Several authors have offered guidelines for the interpretation of a correlation coefficient. However, all such criteria are in some ways arbitrary and should not be observed too strictly. The interpretation of a correlation coefficient depends on the context and purposes. A correlation of 0.8 may be very low if one is verifying a physical law using high-quality instruments, but may be regarded as very high in the social sciences where there may be a greater contribution from complicating factors.

Inference

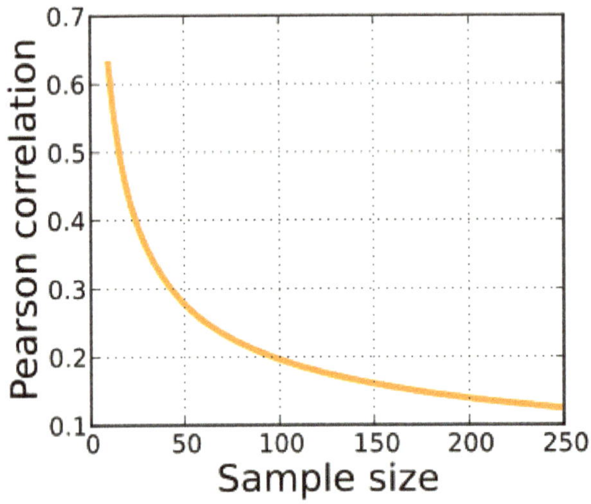

A graph showing the minimum absolute value of Pearson's correlation coefficient that is significantly different from zero at the 0.05 level, for a given sample size.

Statistical inference based on Pearson's correlation coefficient often focuses on one of the following two aims:

- One aim is to test the null hypothesis that the true correlation coefficient ρ is equal to 0, based on the value of the sample correlation coefficient r.
- The other aim is to derive a confidence interval that, on repeated sampling, has a given probability of containing ρ.

We discuss methods of achieving one or both of these aims below.

Using a Permutation Test

Permutation tests provide a direct approach to performing hypothesis tests and constructing confidence intervals. A permutation test for Pearson's correlation coefficient involves the following two steps:

- Using the original paired data (x_i, y_i), randomly redefine the pairs to create a new data set (x_i, $y_{i'}$), where the i′ are a permutation of the set {1,...,n}. The permutation i′ is selected randomly, with equal probabilities placed on all n! possible permutations. This is equivalent to drawing the i′ randomly "without replacement" from the set {1, ..., n}. A closely related and equally justified (bootstrapping) approach is to separately draw the i and the i′ "with replacement" from {1, ..., n};
- Construct a correlation coefficient r from the randomized data.

To perform the permutation test, repeat steps (1) and (2) a large number of times. The p-value for the permutation test is the proportion of the r values generated in step (2) that are larger than the Pearson correlation coefficient that was calculated from the original data. Here "larger" can mean either that the value is larger in magnitude, or larger in signed value, depending on whether a two-sided or one-sided test is desired.

Using a Bootstrap

The bootstrap can be used to construct confidence intervals for Pearson's correlation coefficient. In the "non-parametric" bootstrap, n pairs (x_i, y_i) are resampled "with replacement" from the observed set of n pairs, and the correlation coefficient r is calculated based on the resampled data. This process is repeated a large number of times, and the empirical distribution of the resampled r values are used to approximate the sampling distribution of the statistic. A 95% confidence interval for ρ can be defined as the interval spanning from the 2.5th to the 97.5th percentile of the resampled r values.

Testing using Student's t-Distribution

For pairs from an uncorrelated bivariate normal distribution, the sampling distribution of a certain function of Pearson's correlation coefficient follows Student's t-distribution with degrees of freedom n − 2. Specifically, if the underlying variables have a bivariate normal distribution, the variable

$$t = r\sqrt{\frac{n-2}{1-r^2}}$$

has a Student's t-distribution in the null case (zero correlation). This also holds approximately even if the observed values are non-normal, provided sample sizes are not very small. For determining the critical values for r the inverse of this transformation is also needed:

$$r = \frac{t}{\sqrt{n-2+t^2}}.$$

Alternatively, large sample approaches can be used.

Early work on the distribution of the sample correlation coefficient was carried out by Anil Kumar Gain and R. A. Fisher

Another early paper provides graphs and tables for general values of ρ, for small sample sizes, and discusses computational approaches.

Using the Exact Distribution

For data that follows a bivariate normal distribution, the exact density function f(r) for the sample correlation coefficient r of a normal bivariate is

$$f(r) = \frac{(n-2)\,\mathbf{\Gamma}(n-1)(1-\rho^2)^{\frac{n-1}{2}}(1-r^2)^{\frac{n-4}{2}}}{\sqrt{2\pi}\,\mathbf{\Gamma}\left(n-\frac{1}{2}\right)(1-\rho r)^{n-\frac{3}{2}}}\,{}_2\mathbf{F}_1\left(\frac{1}{2},\frac{1}{2};\frac{2n-1}{2};\frac{\rho r+1}{2}\right)$$

where:

- Γ is the gamma function,
- ${}_2\mathbf{F}_1(a,b;c;z)$ is the Gaussian hypergeometric function.

In the special case when $\rho = 0$, the exact density function f(r) can be written as:

$$f(r) = \frac{(1-r^2)^{\frac{n-4}{2}}}{\mathbf{B}\left(\frac{1}{2}, \frac{n-2}{2}\right)},$$

where:

- B is the beta function, which is one way of writing the density of a Student's t-distribution, as above.

Using the Fisher Transformation

In practice, confidence intervals and hypothesis tests relating to ρ are usually carried out using the Fisher transformation:

$$F(r) = \frac{1}{2}\ln\frac{1+r}{1-r} = \operatorname{arctanh}(r).$$

If F(r) is the Fisher transformation of r, and n is the sample size, then F(r) approximately follows a normal distribution with

$$\text{mean} = F(\rho) = \operatorname{arctanh}(\rho) \quad \text{and standard error} \quad \text{SE} = \frac{1}{\sqrt{n-3}}.$$

Thus, a z-score is

$$z = \frac{x - \text{mean}}{\text{SE}} = [F(r) - F(\rho_0)]\sqrt{n-3}$$

under the null hypothesis of that $\rho = \rho_0$, given the assumption that the sample pairs are independent and identically distributed and follow a bivariate normal distribution. Thus an approximate p-value can be obtained from a normal probability table. For example, if z = 2.2 is observed and a two-sided p-value is desired to test the null hypothesis that $\rho = 0$, the p-value is 2·Φ(−2.2) = 0.028, where Φ is the standard normal cumulative distribution function.

To obtain a confidence interval for ρ, we first compute a confidence interval for $F(\rho)$:

$$100(1-\alpha)\%\text{CI}: \operatorname{arctanh}(\rho) \in [\operatorname{arctanh}(r) \pm z_{\alpha/2}SE]$$

The inverse Fisher transformation bring the interval back to the correlation scale.

$$100(1-\alpha)\%\text{CI}: \rho \in [\tanh(\operatorname{arctanh}(r) - z_{\alpha/2}SE), \tanh(\operatorname{arctanh}(r) + z_{\alpha/2}SE)]$$

For example, suppose we observe r = 0.3 with a sample size of n=50, and we wish to obtain a 95% confidence interval for ρ. The transformed value is arctanh(r) = 0.30952, so the confidence interval on the transformed scale is 0.30952 ± 1.96/√47, or (0.023624, 0.595415). Converting back to the correlation scale yields (0.024, 0.534).

Pearson's Correlation and Least Squares Regression Analysis

The square of the sample correlation coefficient is typically denoted r^2 and is a special case of the coefficient of determination. In this case, it estimates the fraction of the variance in Y that is explained by X in a simple linear regression. So if we have the observed dataset $\{y_1,...y_n\}$ and the fitted dataset $\{f_1,...f_n\}$, and we denote the fitted dataset $\{f_1,...f_n\}$ with $\{\hat{y}_1,...\hat{y}_n\}$, then as a starting point the total variation in the Y_i around their average value can be decomposed as follows

$$\sum_i (Y_i - \bar{Y})^2 = \sum_i (Y_i - \hat{Y}_i)^2 + \sum_i (\hat{Y}_i - \bar{Y})^2,$$

where the $\hat{Y}_i$ are the fitted values from the regression analysis. This can be rearranged to give

$$1 = \frac{\sum_i (Y_i - \hat{Y}_i)^2}{\sum_i (Y_i - \bar{Y})^2} + \frac{\sum_i (\hat{Y}_i - \bar{Y})^2}{\sum_i (Y_i - \bar{Y})^2}.$$

The two summands above are the fraction of variance in Y that is explained by X (right) and that is unexplained by X (left).

Next, we apply a property of least square regression models, that the sample covariance between $\{f_1,...f_n\}$ and $\{f_1,...f_n\}$ with $\{\hat{y}_1,...\hat{y}_n\}$ is zero. Thus, the sample correlation coefficient between the observed and fitted response values in the regression can be written (calculation is under expectation, assumes Gaussian statistics)

$$\begin{aligned} r(Y,\hat{Y}) &= \frac{\sum_i (Y_i - \bar{Y})(\hat{Y}_i - \bar{Y})}{\sqrt{\sum_i (Y_i - \bar{Y})^2 \cdot \sum_i (\hat{Y}_i - \bar{Y})^2}} \\ &= \frac{\sum_i (Y_i - \hat{Y}_i + \hat{Y}_i - \bar{Y})(\hat{Y}_i - \bar{Y})}{\sqrt{\sum_i (Y_i - \bar{Y})^2 \cdot \sum_i (\hat{Y}_i - \bar{Y})^2}} \\ &= \frac{\sum_i [(Y_i - \hat{Y}_i)(\hat{Y}_i - \bar{Y}) + (\hat{Y}_i - \bar{Y})^2]}{\sqrt{\sum_i (Y_i - \bar{Y})^2 \cdot \sum_i (\hat{Y}_i - \bar{Y})^2}} \\ &= \frac{\sum_i (\hat{Y}_i - \bar{Y})^2}{\sqrt{\sum_i (Y_i - \bar{Y})^2 \cdot \sum_i (\hat{Y}_i - \bar{Y})^2}} \\ &= \sqrt{\frac{\sum_i (\hat{Y}_i - \bar{Y})^2}{\sum_i (Y_i - \bar{Y})^2}}. \end{aligned}$$

Thus

$$r(Y,\hat{Y})^2 = \frac{\sum_i (\hat{Y}_i - \bar{Y})^2}{\sum_i (Y_i - \bar{Y})^2}$$

where

- $r(Y,\hat{Y})^2$ is the proportion of variance in Y explained by a linear function of X.

That equation can be written as:

$$r(Y,\hat{Y})^2 = \frac{SS_{\text{reg}}}{SS_{\text{tot}}}$$

where

- SS_{reg} is the regression sum of squares, also called the explained sum of squares
- SS_{tot} is the total sum of squares (proportional to the variance of the data)
- $SS_{\text{reg}} = \sum_i (\hat{Y}_i - \bar{Y})^2$
- $SS_{\text{tot}} = \sum_i (Y_i - \bar{Y})^2$

Sensitivity to the Data Distribution

Existence

The population Pearson correlation coefficient is defined in terms of moments, and therefore exists for any bivariate probability distribution for which the population covariance is defined and the marginal population variances are defined and are non-zero. Some probability distributions such as the Cauchy distribution have undefined variance and hence ρ is not defined if X or Y follows such a distribution. In some practical applications, such as those involving data suspected to follow a heavy-tailed distribution, this is an important consideration. However, the existence of the correlation coefficient is usually not a concern; for instance, if the range of the distribution is bounded, ρ is always defined.

Sample Size

- If the sample size is moderate or large and the population is normal then in the case of the bivariate normal distribution, the sample correlation coefficient is the maximum likelihood estimate of the population correlation coefficient, and is asymptotically unbiased and efficient, which roughly means that it is impossible to construct a more accurate estimate than the sample correlation coefficient.
- If the sample size is large and the population is not normal, then the sample correlation coefficient remains approximately unbiased, but may not be efficient.

- If the sample size is large then sample correlation coefficient is a consistent estimator of the population correlation coefficient as long as the sample means, variances, and covariance are consistent (which is guaranteed when the law of large numbers can be applied).
- If the sample size is small then the sample correlation coefficient r is not an unbiased estimate of ρ. The adjusted correlation coefficient must be used instead.

Robustness

Like many commonly used statistics, the sample statistic r is not robust, so its value can be misleading if outliers are present. Specifically, the PMCC is neither distributionally robust, nor outlier resistant. Inspection of the scatterplot between X and Y will typically reveal a situation where lack of robustness might be an issue, and in such cases it may be advisable to use a robust measure of association. Note however that while most robust estimators of association measure statistical dependence in some way, they are generally not interpretable on the same scale as the Pearson correlation coefficient.

Statistical inference for Pearson's correlation coefficient is sensitive to the data distribution. Exact tests, and asymptotic tests based on the Fisher transformation can be applied if the data are approximately normally distributed, but may be misleading otherwise. In some situations, the bootstrap can be applied to construct confidence intervals, and permutation tests can be applied to carry out hypothesis tests. These non-parametric approaches may give more meaningful results in some situations where bivariate normality does not hold. However the standard versions of these approaches rely on exchangeability of the data, meaning that there is no ordering or grouping of the data pairs being analyzed that might affect the behavior of the correlation estimate.

A stratified analysis is one way to either accommodate a lack of bivariate normality, or to isolate the correlation resulting from one factor while controlling for another. If W represents cluster membership or another factor that it is desirable to control, we can stratify the data based on the value of W, then calculate a correlation coefficient within each stratum. The stratum-level estimates can then be combined to estimate the overall correlation while controlling for W.

Variants

Variations of the correlation coefficient can be calculated for different purposes. Here are some examples.

Adjusted Correlation Coefficient

The sample correlation coefficient r is not an unbiased estimate of ρ. For data that follows a bivariate normal distribution, the expectation E(r) for the sample correlation coefficient r of a normal bivariate is

$$\mathbf{E}[r] = \rho - \frac{\rho\left(1-\rho^2\right)}{2n} + \cdots,$$ therefore r is a biased estimator of ρ.

The unique minimum variance unbiased estimator r_{adj} is given by

$$(1) \qquad r_{adj} = r\,{}_2\mathbf{F}_1\left(\frac{1}{2}, \frac{1}{2}; \frac{n-1}{2}; 1 - r^2\right),$$

where:

- r,n are defined as above,
- ${}_2\mathbf{F}_1(a, b; c; z)$ is the Gaussian hypergeometric function.

An approximately unbiased estimator r_{adj} can be obtained by truncating E[r] and solving this truncated equation:

$$(2) \qquad r = \mathbf{E}[r] = r_{adj} - \frac{r_{adj}\left(1 - r_{adj}^2\right)}{2n}.$$

The solution to equation (2) is:

$$(3) \qquad r_{adj} = r\left[1 + \frac{1 - r^2}{2n}\right],$$

where in (3):

- r, n are defined as above,
- r_{adj} is a suboptimal estimator,
- r_{adj} can also be obtained by maximizing log(f(r)),
- r_{adj} has minimum variance for large values of n,
- r_{adj} has a bias of order 1/(n-1).

Another proposed adjusted correlation coefficient is:

$$r_{adj} = \sqrt{1 - \frac{(1 - r^2)(n - 1)}{(n - 2)}}.$$

Note that $r_{adj} \approx r$ for large values of n.

Weighted Correlation Coefficient

Suppose observations to be correlated have differing degrees of importance that can be expressed with a weight vector w. To calculate the correlation between vectors x and y with the weight vector w (all of length n),

- Weighted mean:

$$\mathrm{m}(x; w) = \frac{\sum_i w_i x_i}{\sum_i w_i}.$$

- Weighted covariance

$$\operatorname{cov}(x,y;w) = \frac{\sum_i w_i (x_i - \operatorname{m}(x;w))(y_i - \operatorname{m}(y;w))}{\sum_i w_i}.$$

- Weighted correlation

$$\operatorname{corr}(x,y;w) = \frac{\operatorname{cov}(x,y;w)}{\sqrt{\operatorname{cov}(x,x;w)\operatorname{cov}(y,y;w)}}.$$

Reflective Correlation Coefficient

The reflective correlation is a variant of Pearson's correlation in which the data are not centered around their mean values. The population reflective correlation is

$$\operatorname{Corr}_r(X,Y) = \frac{E[XY]}{\sqrt{EX^2 \cdot EY^2}}.$$

The reflective correlation is symmetric, but it is not invariant under translation:

$$\operatorname{Corr}_r(X,Y) = \operatorname{Corr}_r(Y,X) = \operatorname{Corr}_r(X,bY) \neq \operatorname{Corr}_r(X,a+bY), \quad a \neq 0, b > 0.$$

The sample reflective correlation is

$$rr_{xy} = \frac{\sum x_i y_i}{\sqrt{(\sum x_i^2)(\sum y_i^2)}}.$$

The weighted version of the sample reflective correlation is

$$rr_{xy,w} = \frac{\sum w_i x_i y_i}{\sqrt{(\sum w_i x_i^2)(\sum w_i y_i^2)}}.$$

Scaled Correlation Coefficient

Scaled correlation is a variant of Pearson's correlation in which the range of the data is restricted intentionally and in a controlled manner to reveal correlations between fast components in time series. Scaled correlation is defined as average correlation across short segments of data.

Let K be the number of segments that can fit into the total length of the signal T for a given scale s:

$$K = \operatorname{round}\left(\frac{T}{s}\right).$$

The scaled correlation across the entire signals $\bar{r}_s$ is then computed as

$$\bar{r}_s = \frac{1}{K}\sum_{k=1}^{K} r_k,$$

where r_k is Pearson's coefficient of correlation for segment k.

By choosing the parameter s, the range of values is reduced and the correlations on long time scale are filtered out, only the correlations on short time scales being revealed. Thus, the contributions of slow components are removed and those of fast components are retained.

Pearson's Distance

A distance metric for two variables X and Y known as Pearson's distance can be defined from their correlation coefficient as

$$d_{X,Y} = 1 - \rho_{X,Y}.$$

Considering that the Pearson correlation coefficient falls between [−1, 1], the Pearson distance lies in [0, 2].

Circular Correlation Coefficient

For variables X = $\{x_1,...,x_n\}$ and Y = $\{y_1,...,y_n\}$ that are defined on the unit circle [0, 2π), it is possible to define a circular analog of Pearson›s coefficient. This is done by transforming data points in X and Y with a sine function such that the correlation coefficient is given as:

$$r_{circular} = \frac{\sum_{i=1}^{n} \sin(x_i - \bar{x}) \sin(y_i - \bar{y})}{\sqrt{\sum_{i=1}^{n} \sin(x_i - \bar{x})^2}\sqrt{\sum_{i=1}^{n} \sin(y_i - \bar{y})^2}}$$

where $\bar{x}$ and $\bar{y}$ are the circular means of X and Y. This measure can be useful in fields like meteorology where the angular direction of data is important.

Heavy Noise Conditions

Under heavy noise conditions, extracting the correlation coefficient between two sets of stochastic variables is nontrivial, in particular where Canonical Correlation Analysis reports on degraded correlation values due to the heavy noise contributions. A generalization of the approach is given elsewhere.

Missing Data

In case of missing data, Garren derived the maximum likelihood estimator.

Removing Correlation

It is always possible to remove the correlation between random variables with a linear transformation, even if the relationship between the variables is nonlinear. A presentation of this result for population distributions is given by Cox & Hinkley.

A corresponding result exists for sample correlations, in which the sample correlation is reduced to zero. Suppose a vector of n random variables is sampled m times. Let X be a matrix where $X_{i,j}$ is the jth variable of sample i. Let $Z_{m,m}$ be an m by m square matrix with every element 1. Then D is the data transformed so every random variable has zero mean, and T is the data transformed so all variables have zero mean and zero correlation with all other variables – the sample covariance matrix of T will be the identity matrix. This has to be further divided by the standard deviation to get unit variance. The transformed variables will be uncorrelated, even though they may not be independent.

$$D = X - \frac{1}{m} Z_{m,m} X$$

$$T = D(D^T D)^{-\frac{1}{2}},$$

where an exponent of −1/2 represents the matrix square root of the inverse of a matrix. The covariance matrix of T will be the identity matrix. If a new data sample x is a row vector of n elements, then the same transform can be applied to x to get the transformed vectors d and t:

$$d = x - \frac{1}{m} Z_{1,m} X,$$

$$t = d(D^T D)^{-\frac{1}{2}}.$$

This decorrelation is related to principal components analysis for multivariate data.

CONFOUNDING

In statistics, a confounding variable (also confounding factor, a confound, a lurking variable or a confounder) is an extraneous variable in a statistical model that correlates (directly or inversely) with both the dependent variable and the independent variable, in a way that "explains away" some or all of the correlation between these two variables.

While specific definitions may vary, in essence a confounding variable fits the following four criteria, here given in a hypothetical situation with variable of interest "V", confounding variable "C" and outcome of interest "O":

1. C is associated (inversely or directly) with O
2. C is associated with O, independent of V
3. C is associated (inversely or directly) with V

4. C is not in the causal pathway of V to O (C is not a direct consequence of V, not a way by which V produces O)

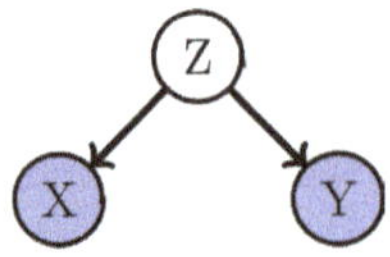

Illustration of a simple confounding case: in this graphical model, given Z, there is no association between X and Y. However, not observing Z will create spurious association between X and Y. In the latter case, Z is called a confounding factor.

The preceding correlation-based definition, however, is metaphorical at best – a growing number of analysts agree that confounding is a causal concept, and as such, cannot be described in terms of correlations nor associations.

Causal Definition

The concept of confounding must be defined, and managed, in terms of the data generating model (as in the Figure above). Specifically, let X be some independent variable, Y some dependent variable. To estimate the effect of X on Y, the statistician must suppress the effects of extraneous variables that influence both X and Y. We say that, X and Y are confounded by some other variable Z whenever Z is a cause of both X and Y.

In the causal framework, denote $P(y|do(x))$ as the probability of event Y = y under the hypothetical intervention X = x. X and Y are not confounded if and only if the following holds:

$$P(y|do(x)) = P(y|x)$$

for all values X = x and Y = y, where $P(y|x)$ is the conditional probability upon seeing X = x. Intuitively, this equality states that X and Y are not confounded whenever the observationally witnessed association between them is the same as the association that would be measured in a controlled experiment, with x randomized.

In principle, the defining equality P(y|do(x)) = P(y|x) can be verified from the data generating model assuming we have all the equations and probabilities associated with the model. This is done by simulating an intervention do(X = x) and checking whether the resulting probability of Y equals the conditional probability P(y|x). It turns out, however, that graph structure alone is sufficient for verifying the equality P(y|do(x)) = P(y|x).

Control of Confounding

Consider a researcher attempting to assess the effectiveness of drug X, from population data in which drug usage was a patient's choice. Data show that gender(Z) differences influence a patient's choice of drug as well as their chances of recovery (Y). In this scenario, gender Z confounds the relation between X and Y since Z is a cause of both X and Y:

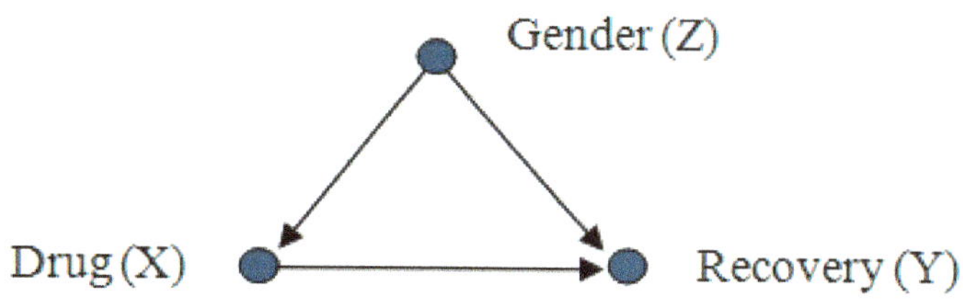

Indeed, we will encounter the inequality:

$$P(y|do(x)) \neq P(y|x)$$

because the observational quantity contains information about the correlation between X and Z, and the interventional quantity does not (since X is not correlated with Z in a randomized experiment). Clearly the statistician desires the unbiased estimate $P(y|do(x))$, but in cases where only observational data are available, an unbiased estimate can only be obtained by "adjusting" for all confounding factors, namely, conditioning on their various values and averaging the result. In the case of a single confounder Z, this leads to the "adjustment formula":

$$P(y|do(x)) = \sum_z P(y|x,z)P(z)$$

which gives an unbiased estimate for the causal effect of X on Y. The same adjustment formula works when there are multiple confounders except, in this case, the choice of a set Z of variables that would guarantee unbiased estimates must be done with caution. The criterion for a proper choice of variables is called the Back-Door and requires that the chosen set Z "blocks" (or intercepts) every path from X to Y that ends with an arrow into X. Such sets are called "Back-Door admissible" and may include variables which are not common causes of X and Y, but merely proxies thereof.

Returning to the drug use example, since Z complies with the Back-Door requirement (i.e., it intercepts the one Back-Door path $X \leftarrow Z \rightarrow Y$), the Back-Door adjustment formula is valid:

$$\begin{aligned} P(Y = \text{recovered} \mid do(x = \text{give drug})) = {} & P(Y = \text{recovered} \mid X = \text{give drug}, Z = \text{male})P(Z = \text{male}) \\ & + P(Y = \text{recovered} \mid X = \text{give drug}, Z = \text{female})P(Z = \text{female}) \end{aligned}$$

In this way the physician can predict the likely effect of administering the drug from observational studies in which the conditional probabilities appearing on the right-hand side of the equation can be estimated by regression.

Contrary to common beliefs, adding covariates to the adjustment set Z can introduce bias. A typical counterexample occurs when Z is a common effect of X and Y, a case in which Z is not a confounder (i.e., the null set is Back-door admissible) and adjusting for Z would create bias known as "collider bias" or "Berkson's paradox." In general, confounding can be controlled by adjustment if and only if there is a set of observed covariates that satisfies the Back-Door condition. Moreover, if Z is such a set, then the adjustment formula of Eq. (3) is valid <4,5>. Pearl's do-calculus provide additional conditions under which P(y|do(x)) can be estimated, not necessarily by adjustment.

History

According to Morabia (2011), the word derives from the Medieval Latin verb "confudere", which meant "mixing", and was probably chosen to represent the confusion between the cause one wishes to assess and other causes that may affect the outcome and thus confuse, or stand in the way of the desired assessment. Fisher used the word "confounding" in his 1935 book "The Design of Experiments" to denote any source of error in his ideal of randomized experiment. According to Vandenbroucke (2004) it was Kish who used the word "confounding" in the modern sense of the word, to mean "incomparability" of two or more groups (e.g., exposed and unexposed) in an observational study.

Formal conditions defining what makes certain groups "comparable" and others "incomparable" were later developed in epidemiology by Greenland and Robins (1986) using the counterfactual language of Neyman (1935) and Rubin (1974). These were later supplemented by graphical criteria such as the Back-Door condition (Pearl 1993; Greenland, Pearl and Robins, 1999).

Graphical criteria were shown to be formally equivalent to the counterfactual definition, but more transparent to researchers relying on process models.

Types of Confounding

In the case of risk assessments evaluating the magnitude and nature of risk to human health, it is important to control for confounding to isolate the effect of a particular hazard such as a food additive, pesticide, or new drug. For prospective studies, it is difficult to recruit and screen for volunteers with the same background (age, diet, education, geography, etc.), and in historical studies, there can be similar variability. Due to the inability to control for variability of volunteers and human studies, confounding is a particular challenge. For these reasons, experiments offer a way to avoid most forms of confounding.

In some disciplines, confounding is categorized into different types. In epidemiology, one type is "confounding by indication", which relates to confounding from observational studies. Because prognostic factors may influence treatment decisions (and bias estimates of treatment effects), controlling for known prognostic factors may reduce this problem, but it is always possible that a forgotten or unknown factor was not included or that factors interact complexly. Confounding by indication has been described as the most important limitation of observational studies. Randomized trials are not affected by confounding by indication due to random assignment.

Confounding variables may also be categorised according to their source. The choice of measurement instrument (operational confound), situational characteristics (procedural confound), or inter-individual differences (person confound).

- An operational confounding can occur in both experimental and non-experimental research designs. This type of confounding occurs when a measure designed to assess a particular construct inadvertently measures something else as well.
- A procedural confounding can occur in a laboratory experiment or a quasi-experiment. This type of confound occurs when the researcher mistakenly allows another variable to change along with the manipulated independent variable.
- A person confounding occurs when two or more groups of units are analyzed together (e.g., workers from different occupations), despite varying according to one or more other (observed or unobserved) characteristics (e.g., gender).

Examples

As an example, suppose that there is a statistical relationship between ice-cream consumption and number of drowning deaths for a given period. These two variables have a positive correlation with each other. An evaluator might attempt to explain this correlation by inferring a causal relationship between the two variables (either that ice-cream causes drowning, or that drowning causes ice-cream consumption). However, a more likely explanation is that the relationship between ice-cream consumption and drowning is spurious and that a third, confounding, variable (the season) influences both variables: during the summer, warmer temperatures lead to increased ice-cream consumption as well as more people swimming and thus more drowning deaths.

In another concrete example, say one is studying the relation between birth order (1st child, 2nd child, etc.) and the presence of Down's Syndrome in the child. In this scenario, maternal age would be a confounding variable:

1. Higher maternal age is directly associated with Down's Syndrome in the child
2. Higher maternal age is directly associated with Down's Syndrome, regardless of birth order (a mother having her 1st vs 3rd child at age 50 confers the same risk)
3. Maternal age is directly associated with birth order (the 2nd child, except in the case of twins, is born when the mother is older than she was for the birth of the 1st child)
4. Maternal age is not a consequence of birth order (having a 2nd child does not change the mother's age)

In risk assessments, factors such as age, gender, and educational levels often affect health status and so should be controlled. Beyond these factors, researchers may not consider or have access to data on other causal factors. An example is on the study of smoking tobacco on human health. Smoking, drinking alcohol, and diet are lifestyle activities that are related. A risk assessment that looks at the effects of smoking but does not control for alcohol consumption or diet may overestimate the risk of smoking. Smoking and confounding are reviewed in occupational risk assessments such as the safety of coal mining. When there is not a large sample population of non-smokers or non-drinkers in a particular occupation, the risk assessment may be biased towards finding a negative effect on health.

Decreasing the Potential for Confounding

A reduction in the potential for the occurrence and effect of confounding factors can be obtained by increasing the types and numbers of comparisons performed in an analysis. If measures or manipulations of core constructs are confounded (i.e. operational or procedural confounds exist), subgroup analysis may not reveal problems in the analysis. Additionally, increasing the number of comparisons can create other problems.

Peer review is a process that can assist in reducing instances of confounding, either before study implementation or after analysis has occurred. Peer review relies on collective expertise within a discipline to identify potential weaknesses in study design and analysis, including ways in which results may depend on confounding. Similarly, replication can test for the robustness of findings from one study under alternative study conditions or alternative analyses (e.g., controlling for potential confounds not identified in the initial study).

Confounding effects may be less likely to occur and act similarly at multiple times and locations. In selecting study sites, the environment can be characterized in detail at the study sites to ensure sites are ecologically similar and therefore less likely to have confounding variables. Lastly, the relationship between the environmental variables that possibly confound the analysis and the measured parameters can be studied. The information pertaining to environmental variables can then be used in site-specific models to identify residual variance that may be due to real effects.

Depending on the type of study design in place, there are various ways to modify that design to actively exclude or control confounding variables:

- Case-control studies assign confounders to both groups, cases and controls, equally. For example if somebody wanted to study the cause of myocardial infarct and thinks that the age is a probable confounding variable, each 67-year-old infarct patient will be matched with a healthy 67-year-old "control" person. In case-control studies, matched variables most often are the age and sex. Drawback: Case-control studies are feasible only when it is easy to find controls, i.e. persons whose status vis-à-vis all known potential confounding factors is the same as that of the case's patient: Suppose a case-control study attempts to find the cause of a given disease in a person who is 1) 45 years old, 2) African-American, 3) from Alaska, 4) an avid football player, 5) vegetarian, and 6) working in education. A theoretically perfect control would be a person who, in addition to not having the disease being investigated, matches all these characteristics and has no diseases that the patient does not also have — but finding such a control would be an enormous task.
- Cohort studies: A degree of matching is also possible and it is often done by only admitting certain age groups or a certain sex into the study population, creating a cohort of people who share similar characteristics and thus all cohorts are comparable in regard to the possible confounding variable. For example, if age and sex are thought to be confounders, only 40 to 50 years old males would be involved in a cohort study that would assess the myocardial infarct risk in cohorts that either are physically active or inactive. Drawback: In cohort studies,

the overexclusion of input data may lead researchers to define too narrowly the set of similarly situated persons for whom they claim the study to be useful, such that other persons to whom the causal relationship does in fact apply may lose the opportunity to benefit from the study's recommendations. Similarly, "over-stratification" of input data within a study may reduce the sample size in a given stratum to the point where generalizations drawn by observing the members of that stratum alone are not statistically significant.

- Double blinding: conceals from the trial population and the observers the experiment group membership of the participants. By preventing the participants from knowing if they are receiving treatment or not, the placebo effect should be the same for the control and treatment groups. By preventing the observers from knowing of their membership, there should be no bias from researchers treating the groups differently or from interpreting the outcomes differently.
- Randomized controlled trial: A method where the study population is divided randomly in order to mitigate the chances of self-selection by participants or bias by the study designers. Before the experiment begins, the testers will assign the members of the participant pool to their groups (control, intervention, parallel), using a randomization process such as the use of a random number generator. For example, in a study on the effects of exercise, the conclusions would be less valid if participants were given a choice if they wanted to belong to the control group which would not exercise or the intervention group which would be willing to take part in an exercise program. The study would then capture other variables besides exercise, such as pre-experiment health levels and motivation to adopt healthy activities. From the observer's side, the experimenter may choose candidates who are more likely to show the results the study wants to see or may interpret subjective results (more energetic, positive attitude) in a way favorable to their desires.
- Stratification: As in the example above, physical activity is thought to be a behavior that protects from myocardial infarct; and age is assumed to be a possible confounder. The data sampled is then stratified by age group – this means that the association between activity and infarct would be analyzed per each age group. If the different age groups (or age strata) yield much different risk ratios, age must be viewed as a confounding variable. There exist statistical tools, among them Mantel–Haenszel methods, that account for stratification of data sets.
- Controlling for confounding by measuring the known confounders and including them as covariates is multivariable analysis such as regression analysis. Multivariate analyses reveal much less information about the strength or polarity of the confounding variable than do stratification methods. For example, if multivariate analysis controls for antidepressant, and it does not stratify antidepressants for TCA and SSRI, then it will ignore that these two classes of antidepressant have opposite effects on myocardial infarction, and one is much stronger than the other.

All these methods have their drawbacks:

- The best available defense against the possibility of spurious results due to confounding is often to dispense with efforts at stratification and instead conduct a randomized study of a sufficiently large sample taken as a whole, such that all potential confounding variables (known and unknown) will be distributed by chance across all study groups and hence will be uncorrelated with the binary variable for inclusion/exclusion in any group.
- Ethical considerations: In double blind and randomized controlled trials, participants are not aware that they are recipients of sham treatments and may be denied effective treatments. There is resistance to randomized controlled trials in surgery because patients would agree to invasive surgery which carry risks under the understanding that they were receiving treatment.

COEFFICIENT OF DETERMINATION

In statistics, the coefficient of determination, denoted R^2 or r^2 and pronounced "R squared", is a number that indicates the proportion of the variance in the dependent variable that is predictable from the independent variable.

It is a statistic used in the context of statistical models whose main purpose is either the prediction of future outcomes or the testing of hypotheses, on the basis of other related information. It provides a measure of how well observed outcomes are replicated by the model, based on the proportion of total variation of outcomes explained by the model.

There are several definitions of R^2 that are only sometimes equivalent. One class of such cases includes that of simple linear regression where r^2 is used instead of R^2. When an intercept is included, then r^2 is simply the square of the sample correlation coefficient (i.e., r) between the outcomes and their predicted values. If additional regressors are included, R^2 is the square of the coefficient of multiple correlation. In both such cases, the coefficient of determination ranges from 0 to 1.

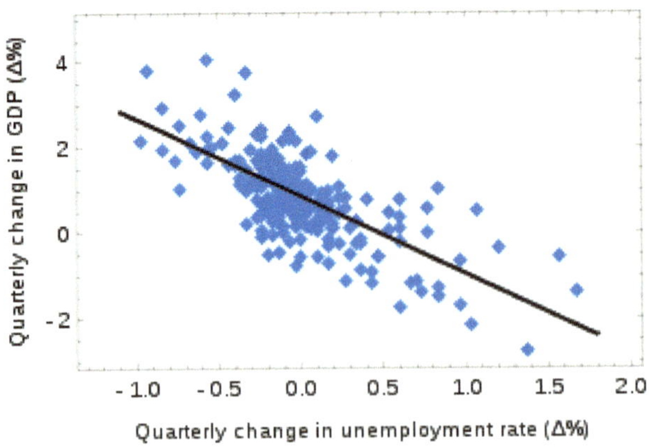

Ordinary least squares regression of Okun's law. Since the regression line does not miss any of the points by very much, the R^2 of the regression is relatively high.

Important cases where the computational definition of R^2 can yield negative values, depending on the definition used, arise where the predictions that are being compared to the corresponding outcomes have not been derived from a model-fitting procedure using those data, and where linear regression is conducted without including an intercept. Additionally, negative values of R^2 may occur when fitting non-linear functions to data. In cases where negative values arise, the mean of the data provides a better fit to the outcomes than do the fitted function values, according to this particular criterion.

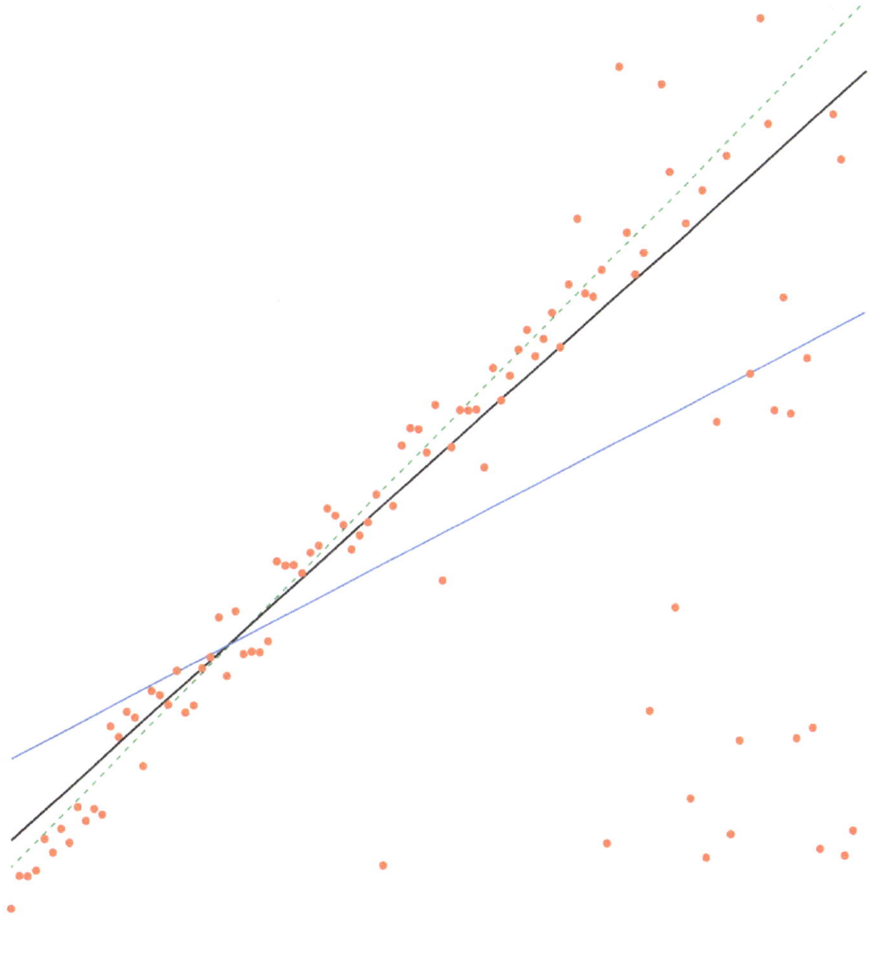

Comparison of the Theil–Sen estimator (black) and simple linear regression (blue) for a set of points with outliers. Because of the many outliers, neither of the regression lines fits the data well, as measured by the fact that neither gives a very high R^2.

Definitions

A data set has n values marked $y_1,...,y_n$ (collectively known as y_i or as a vector $y = [y_1,...,y_n]^T$), each associated with a predicted (or modeled) value $f_1,...,f_n$ (known as f_i, or sometimes $\hat{y}_i$, as a vector f).

Define the residuals as $e_i = y_i - f_i$ (forming a vector e).

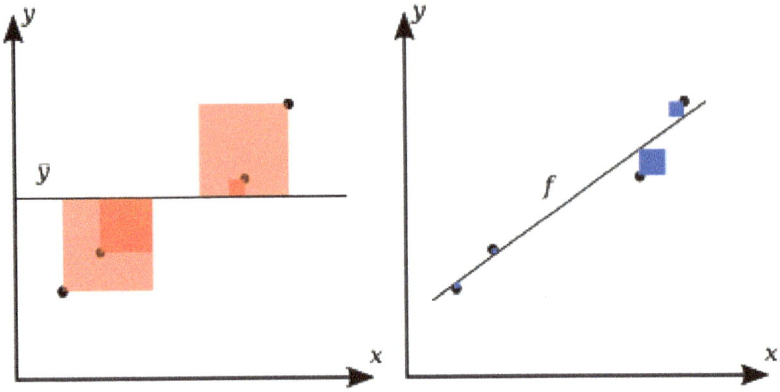

$R^2 = 1 - \frac{SS_{res}}{SS_{tot}}$ The better the linear regression (on the right) fits the data in comparison to the simple average (on the left graph), the closer the value of R^2 is to 1. The areas of the blue squares represent the squared residuals with respect to the linear regression. The areas of the red squares represent the squared residuals with respect to the average value.

If $\bar{y}$ is the mean of the observed data:

$$\bar{y} = \frac{1}{n}\sum_{i=1}^{n} y_i$$

then the variability of the data set can be measured using three sums of squares formulas:

- The total sum of squares (proportional to the variance of the data):

$$SS_{\text{tot}} = \sum_i (y_i - \bar{y})^2,$$

- The regression sum of squares, also called the explained sum of squares:

$$SS_{\text{reg}} = \sum_i (f_i - \bar{y})^2,$$

- The sum of squares of residuals, also called the residual sum of squares:

$$SS_{\text{res}} = \sum_i (y_i - f_i)^2 = \sum_i e_i^2$$

The most general definition of the coefficient of determination is

$$R^2 \equiv 1 - \frac{SS_{\text{res}}}{SS_{\text{tot}}}.$$

Relation to Unexplained Variance

In a general form, R^2 can be seen to be related to the fraction of unexplained variance (FUV), since the second term compares the unexplained variance (variance of the model's errors) with the total variance (of the data):

$$R^2 = 1 - \text{FUV}$$

As Explained Variance

Suppose r = 0.7, meaning $r^2 = 0.49$. This implies that 49% of the variability between the two variables has been accounted for, and the remaining 51% of the variability is still unaccounted for. In some cases the total sum of squares equals the sum of the two other sums of squares defined above,

$$SS_{\text{res}} + SS_{\text{reg}} = SS_{\text{tot}}.$$

See partitioning in the general OLS model for a derivation of this result for one case where the relation holds. When this relation does hold, the above definition of R^2 is equivalent to

$$R^2 = \frac{SS_{\text{reg}}}{SS_{\text{tot}}} = \frac{SS_{\text{reg}}/n}{SS_{\text{tot}}/n}.$$

In this form R^2 is expressed as the ratio of the explained variance (variance of the model's predictions, which is SS_{reg} / n) to the total variance (sample variance of the dependent variable, which is SS_{tot} / n).

This partition of the sum of squares holds for instance when the model values f_i have been obtained by linear regression. A milder sufficient condition reads as follows: The model has the form

$$f_i = \alpha + \beta q_i$$

where the q_i are arbitrary values that may or may not depend on i or on other free parameters (the common choice $q_i = x_i$ is just one special case), and the coefficients α and β are obtained by minimizing the residual sum of squares.

This set of conditions is an important one and it has a number of implications for the properties of the fitted residuals and the modelled values. In particular, under these conditions:

$$\bar{f} = \bar{y}.$$

As Squared Correlation Coefficient

In linear least squares regression with an estimated intercept term, R^2 equals the square of the Pearson correlation coefficient between the observed and modeled (predicted) data values of the dependent variable. Specifically, R^2 equals the squared Pearson correlation coefficient of the dependent and explanatory variable in an univariate linear least squares regression.

Under more general modeling conditions, where the predicted values might be generated from a model different from linear least squares regression, an R^2 value can be calculated as the square of the correlation coefficient between the original and modeled data values. In this case, the value is not directly a measure of how good the modeled values are, but rather a measure of how good a predictor might be constructed from the modeled values (by creating a revised predictor of the form $\alpha + \beta f_i$). According to Everitt (p. 78), this usage is specifically the definition of the term "coefficient of determination": the square of the correlation between two (general) variables.

Interpretation

R^2 is a statistic that will give some information about the goodness of fit of a model. In regression, the R^2 coefficient of determination is a statistical measure of how well the regression line approximates the real data points. An R^2 of 1 indicates that the regression line perfectly fits the data.

Values of R^2 outside the range 0 to 1 can occur where it is used to measure the agreement between observed and modeled values and where the "modeled" values are not obtained by linear regression and depending on which formulation of R^2 is used. If the first formula above is used, values can be less than zero. If the second expression is used, values can be greater than one. Neither formula is defined for the case where $y_1 = \ldots = y_n = \bar{y}$.

In all instances where R^2 is used, the predictors are calculated by ordinary least-squares regression: that is, by minimizing SS_{res}. In this case R^2 increases as we increase the number of variables in the model (R^2 is monotone increasing with the number of variables included—i.e., it will never decrease). This illustrates a drawback to one possible use of R^2, where one might keep adding variables (Kitchen sink regression) to increase the R^2 value. For example, if one is trying to predict the sales of a model of car from the car's gas mileage, price, and engine power, one can include such irrelevant factors as the first letter of the model's name or the height of the lead engineer designing the car because the R^2 will never decrease as variables are added and will probably experience an increase due to chance alone.

This leads to the alternative approach of looking at the adjusted R^2. The explanation of this statistic is almost the same as R^2 but it penalizes the statistic as extra variables are included in the model. For cases other than fitting by ordinary least squares, the R^2 statistic can be calculated as above and may still be a useful measure. If fitting is by weighted least squares or generalized least squares, alternative versions of R^2 can be calculated appropriate to those statistical frameworks, while the "raw" R^2 may still be useful if it is more easily interpreted. Values for R^2 can be calculated for any type of predictive model, which need not have a statistical basis.

While it is often desirable to have high enough R-squared, there are multiple models with low R-squared that might worry an inexperienced researcher. Low R-squared are rather expected in models with presence of other influential factors that can't be measured. For example, R-squared of 10% and less is appropriate for models that study impact of practicing religion on health and life satisfaction.

In a Non-Simple Linear Model

Consider a linear model with more than a single explanatory variable, of the form

$$Y_i = \beta_0 + \sum_{j=1}^{p} \beta_j X_{i,j} + \varepsilon_i,$$

where, for the ith case, Y_i is the response variable, $X_{i,1}, \ldots, X_{i,p}$ are p regressors, and ε_i is a mean zero error term. The quantities $\beta_0, \ldots, \beta_p$ are unknown coefficients, whose values are estimated by least squares. The coefficient of determination R^2 is a measure of the global fit of the model. Specifically, R^2 is an element of [0, 1] and represents the proportion of variability in Y_i that may be attributed to some linear combination of the regressors (explanatory variables) in X.

R^2 is often interpreted as the proportion of response variation "explained" by the regressors in the model. Thus, $R^2 = 1$ indicates that the fitted model explains all variability in y, while $R^2 = 0$ indicates no 'linear' relationship (for straight line regression, this means that the straight line model is a constant line (slope = 0, intercept = $\bar{y}$) between the response variable and regressors). An interior value such as $R^2 = 0.7$ may be interpreted as follows: "Seventy percent of the variance in the response variable can be explained by the explanatory variables. The remaining thirty percent can be attributed to unknown, lurking variables or inherent variability."

A caution that applies to R^2, as to other statistical descriptions of correlation and association is that "correlation does not imply causation." In other words, while correlations may sometimes provide valuable clues in uncovering causal relationships among variables, a non-zero estimated correlation between two variables is not, on its own, evidence that changing the value of one variable would result in changes in the values of other variables. For example, the practice of carrying matches (or a lighter) is correlated with incidence of lung cancer, but carrying matches does not cause cancer (in the standard sense of "cause").

In case of a single regressor, fitted by least squares, R^2 is the square of the Pearson product-moment correlation coefficient relating the regressor and the response variable. More generally, R^2 is the square of the correlation between the constructed predictor and the response variable. With more than one regressor, the R^2 can be referred to as the coefficient of multiple determination.

Inflation of R^2

In least squares regression, R^2 is weakly increasing with increases in the number of regressors in the model. Because increases in the number of regressors increase the value of R^2, R^2 alone cannot be used as a meaningful comparison of models with very different numbers of independent variables. For a meaningful comparison between two models, an F-test can be performed on the residual sum of squares, similar to the F-tests in Granger causality, though this is not always appropriate. As a reminder of this, some authors denote R^2 by R_p^2, where p is the number of columns in X (the number of explanators including the constant).

To demonstrate this property, first recall that the objective of least squares linear regression is:

$$\min_b SS_{\text{res}}(b) \Rightarrow \min_b \sum_i (y_i - X_i b)^2$$

The optimal value of the objective is weakly smaller as additional columns of X are added, by the fact that less constrained minimization leads to an optimal cost which is weakly smaller than more constrained minimization does. Given the previous conclusion and noting that SS_{tot} depends only on y, the non-decreasing property of R^2 follows directly from the definition above.

The intuitive reason that using an additional explanatory variable cannot lower the R^2 is this: Minimizing SS_{res} is equivalent to maximizing R^2. When the extra variable is included, the data always have the option of giving it an estimated coefficient of zero, leaving the predicted values and the R^2 unchanged. The only way that the optimization problem will give a non-zero coefficient is if doing so improves the R^2.

Caveats

R^2 does not indicate whether:

- the independent variables are a cause of the changes in the dependent variable;
- omitted-variable bias exists;
- the correct regression was used;
- the most appropriate set of independent variables has been chosen;
- there is collinearity present in the data on the explanatory variables;
- the model might be improved by using transformed versions of the existing set of independent variables;
- there are enough data points to make a solid conclusion.

Adjusted R^2

The use of an adjusted R^2 (one common notation is $\bar{R}^2$, pronounced "R bar squared"; another is R^2_{adj}) is an attempt to take account of the phenomenon of the R^2automatically and spuriously increasing when extra explanatory variables are added to the model. It is a modification due to Theil of R^2 that adjusts for the number of explanatory terms in a model relative to the number of data points. The adjusted R^2 can be negative, and its value will always be less than or equal to that of R^2. Unlike R^2, the adjusted R^2 increases only when the increase in R^2 (due to the inclusion of a new explanatory variable) is more than one would expect to see by chance. If a set of explanatory variables with a predetermined hierarchy of importance are introduced into a regression one at a time, with the adjusted R^2computed each time, the level at which adjusted R^2 reaches a maximum, and decreases afterward, would be the regression with the ideal combination of having the best fit without excess/unnecessary terms. The adjusted R^2 is defined as

$$\bar{R}^2 = 1 - (1 - R^2)\frac{n-1}{n-p-1} = R^2 - (1 - R^2)\frac{p}{n-p-1}$$

where p is the total number of explanatory variables in the model (not including the constant term), and n is the sample size.

Adjusted R^2 can also be written as

$$\bar{R}^2 = 1 - \frac{SS_{\text{res}}/\text{df}_e}{SS_{\text{tot}}/\text{df}_t}$$

where df_t is the degrees of freedom n– 1 of the estimate of the population variance of the dependent variable, and df_e is the degrees of freedom n – p – 1 of the estimate of the underlying population error variance.

The principle behind the adjusted R^2 statistic can be seen by rewriting the ordinary R^2 as

$$R^2 = 1 - \frac{VAR_{\text{res}}}{VAR_{\text{tot}}}$$

where $\text{VAR}_{\text{res}} = SS_{\text{res}}/n$ and $\text{VAR}_{\text{tot}} = SS_{\text{tot}}/n$ are the sample variances of the estimated residuals and the dependent variable respectively, which can be seen as biased estimates of the population variances of the errors and of the dependent variable. These estimates are replaced by statistically unbiased versions: $\text{VAR}_{\text{res}} = SS_{\text{res}}/(n-p-1)$ and $\text{VAR}_{\text{tot}} = SS_{\text{tot}}/(n-1)$.

Adjusted R^2 does not have the same interpretation as R^2—while R^2 is a measure of fit, adjusted R^2 is instead a comparative measure of suitability of alternative nested sets of explanators. As such, care must be taken in interpreting and reporting this statistic. Adjusted R^2 is particularly useful in the feature selection stage of model building.

Coefficient of Partial Determination

The coefficient of partial determination can be defined as the proportion of variation that cannot be explained in a reduced model, but can be explained by the predictors specified in a full(er) model. This coefficient is used to provide insight into whether or not one or more additional predictors may be useful in a more fully specified regression model.

The calculation for the partial r^2 is relatively straight forward after estimating two models and generating the ANOVA tables for them. The calculation for the partial $r^2$is:

(SSEreduced − SSEfull) / SSEreduced

which is analogous to the usual coefficient of determination

(SST - SSE) / SST.

Generalized R^2

The generalized R^2 was originally proposed by Cox & Snell, and independently by Magee:

$$R^2 = 1 - \left(\frac{L(0)}{L(\hat{\theta})}\right)^{2/n}$$

where L(0) is the likelihood of the model with only the intercept, $L(\hat{\theta})$ is the likelihood of the estimated model (i.e., the model with a given set of parameter estimates) and n is the sample size.

Nagelkerke noted that it had the following properties:

- It's consistent with the classical coefficient of determination when both can be computed;
- Its value is maximized by the maximum likelihood estimation of a model;
- It is asymptotically independent of the sample size;
- The interpretation is the proportion of the variation explained by the model;
- The values are between 0 and 1, with 0 denoting that model does not explain any variation and 1 denoting that it perfectly explains the observed variation;
- It does not have any unit.

However, in the case of a logistic model, where $L(\hat{\theta})$ cannot be greater than 1, R^1 is between 0 and $R^2_{\max} = 1 - (L(0))^{2/n}$: thus, Nagelkerke suggested the possibility to define a scaled R^2 as R^2/R^2_{max}.

Comparison with Norm of Residuals

Occasionally the norm of residuals is used for indicating goodness of fit. This term is calculated as the square-root of the sum of squared residuals:

$$\text{norm of residuals} = \sqrt{SS_{\text{res}}} = \|e\|.$$

Both R^2 and the norm of residuals have their relative merits. For least squares analysis R^2 varies between 0 and 1, with larger numbers indicating better fits and 1 represents a perfect fit. Norm of residuals varies from 0 to infinity with smaller numbers indicating better fits and zero indicating a perfect fit. One advantage and disadvantage of R^2 is the SS_{tot} term acts to normalize the value. If the y_i values are all multiplied by a constant, the norm of residuals will also change by that constant but R^2 will stay the same. As a basic example, for the linear least squares fit to the set of data:

$$x = 1,\ 2,\ 3,\ 4,\ 5$$
$$y = 1.9,\ 3.7,\ 5.8,\ 8.0,\ 9.6$$

$R^2 = 0.998$, and norm of residuals = 0.302. If all values of y are multiplied by 1000 (for example, in an SI prefix change), then R^2 remains the same, but norm of residuals = 302.

Another way to examine goodness of fit would be to examine residuals as a function of x. Other single parameter indicators include the standard deviation of the residuals, or the RMSE of the residuals. These would have values of 0.151 and 0.174 respectively for the above example given that the fit was linear with an unforced intercept.

REFERENCES

1. Croxton, Frederick Emory; Cowden, Dudley Johnstone; Klein, Sidney (1968) Applied General Statistics, Pitman. ISBN 9780273403159 (page 625)
2. Dietrich, Cornelius Frank (1991) Uncertainty, Calibration and Probability: The Statistics of Scientific and Industrial Measurement 2nd Edition, A. Higler. ISBN 9780750300605 (Page 331)
3. Aitken, Alexander Craig (1957) Statistical Mathematics 8th Edition. Oliver & Boyd. ISBN 9780050013007 (Page 95)
4. Rodgers, J. L.; Nicewander, W. A. (1988). "Thirteen ways to look at the correlation coefficient". The American Statistician. 42 (1): 59–66. doi:10.1080/00031305.1988.10475524. JSTOR 2685263.
5. Dowdy, S. and Wearden, S. (1983). "Statistics for Research", Wiley. ISBN 0-471-08602-9 pp 230
6. Francis, DP; Coats AJ; Gibson D (1999). "How high can a correlation coefficient be?". Int J Cardiol. 69 (2): 185–199. doi:10.1016/S0167-5273(99)00028-5.
7. Yule, G.U and Kendall, M.G. (1950), "An Introduction to the Theory of Statistics", 14th Edition (5th Impression 1968). Charles Griffin & Co. pp 258–270
8. Kendall, M. G. (1955) "Rank Correlation Methods", Charles Griffin & Co.
9. Mahdavi Damghani B. (2013). "The Non-Misleading Value of Inferred Correlation: An Introduction to the Cointelation Model". Wilmott Magazine. doi:10.1002/wilm.10252.
10. Székely, G. J. Rizzo; Bakirov, N. K. (2007). "Measuring and testing independence by correlation of distances". Annals of Statistics. 35 (6): 2769–2794. doi:10.1214/009053607000000505.
11. Székely, G. J.; Rizzo, M. L. (2009). "Brownian distance covariance". Annals of Applied Statistics. 3 (4): 1233–1303. doi:10.1214/09-AOAS312.
12. Lopez-Paz D. and Hennig P. and Schölkopf B. (2013). "The Randomized Dependence Coefficient", "Conference on Neural Information Processing Systems" Reprint
13. Thorndike, Robert Ladd (1947). Research problems and techniques (Report No. 3). Washington DC: US Govt. print. off.
14. Nikolić, D; Muresan, RC; Feng, W; Singer, W (2012). "Scaled correlation analysis: a better way to compute a cross-correlogram". European Journal of Neuroscience: 1–21. doi:10.1111/j.1460-9568.2011.07987.x.
15. Aldrich, John (1995). "Correlations Genuine and Spurious in Pearson and Yule". Statistical Science. 10 (4): 364–376. doi:10.1214/ss/1177009870. JSTOR 2246135.

16. Mahdavi Damghani, Babak (2012). “The Misleading Value of Measured Correlation”. Wilmott. 2012 (1): 64–73. doi:10.1002/wilm.10167.
17. Anscombe, Francis J. (1973). “Graphs in statistical analysis”. The American Statistician. 27: 17–21. doi:10.2307/2682899. JSTOR 2682899.
18. Karl Pearson (20 June 1895) “Notes on regression and inheritance in the case of two parents,” Proceedings of the Royal Society of London, 58 : 240–242.
19. Stigler, Stephen M. (1989). “Francis Galton’s Account of the Invention of Correlation”. Statistical Science. 4 (2): 73–79. doi:10.1214/ss/1177012580. JSTOR 2245329.
20. Gayen, A.K. (1951). “The frequency distribution of the product moment correlation coefficient in random samples of any size draw from non-normal universes”. Biometrika. 38: 219–247. doi:10.1093/biomet/38.1-2.219.
21. Fisher, R.A. (1915). “Frequency distribution of the values of the correlation coefficient in samples from an indefinitely large population”. Biometrika. 10 (4): 507–521. doi:10.1093/biomet/10.4.507.
22. Fisher, R.A. (1921). “On the probable error of a coefficient of correlation deduced from a small sample” (PDF). Metron. 1 (4): 3–32. Retrieved 2009-03-25.
23. Real Statistics Using Excel: Correlation: Basic Concepts, retrieved 2015-02-22
24. Schmid Jr., John (December 1947). “The Relationship between the Coefficient of Correlation and the Angle Included between Regression Lines”. The Journal of Educational Research. 41 (4). JSTOR 27528906.
25. Rummel, R. J. (1976). “Understanding Correlation”.
26. Buda, Andrzej; Jarynowski, Andrzej (December 2010). Life time of correlations and its applications. Wydawnictwo Niezależne. pp. 5–21. ISBN 9788391527290.
27. Cohen, J. (1988). Statistical power analysis for the behavioral sciences (2nd ed.)
28. Rahman, N. A. (1968) A Course in Theoretical Statistics, Charles Griffin and Company, 1968
29. Kendall, M. G., Stuart, A. (1973) The Advanced Theory of Statistics, Volume 2: Inference and Relationship, Griffin. ISBN 0-85264-215-6 (Section 31.19)
30. Gain, A. K. (1951). “The frequency distribution of the product moment correlation coefficient in random samples of any size draw from non-normal universes”. Biometrika. 38: 219–247. doi:10.1093/biomet/38.1-2.219.
31. Soper, H. E.; Young, A. W.; Cave, B. M.; Lee, A.; Pearson, K. (1917). “On the distribution of the correlation coefficient in small samples. Appendix II to the papers of “Student” and R. A. Fisher. A co-operative study”. Biometrika. 11: 328–413. doi:10.1093/biomet/11.4.328.
32. Kenney, J. F. and Keeping, E. S., Mathematics of Statistics, Pt. 2, 2nd ed. Princeton, NJ: Van Nostrand, 1951.
33. Correlation Coefficient – Bivariate Normal Distribution
34. Wilcox, Rand R. (2005). Introduction to robust estimation and hypothesis testing. Academic Press.

35. Devlin, Susan J; Gnanadesikan, R; Kettenring J.R. (1975). "Robust Estimation and Outlier Detection with Correlation Coefficients". Biometrika. 62 (3): 531–545. doi:10.1093/biomet/62.3.531. JSTOR 2335508.
36. Huber, Peter. J. (2004). Robust Statistics. Wiley.
37. Katz., Mitchell H. (2006) Multivariable Analysis – A Practical Guide for Clinicians. 2nd Edition. Cambridge University Press. ISBN 978-0-521-54985-1. ISBN 0-521-54985-X doi:10.2277/052154985X
38. Hotelling, H. (1953). "New Light on the Correlation Coefficient and its Transforms". Journal of the Royal Statistical Society. Series B (Methodological). 15 (2): 193–232. JSTOR 2983768.
39. Olkin, Ingram; Pratt,John W. (March 1958). "Unbiased Estimation of Certain Correlation Coefficients". The Annals of Mathematical Statistics. 29 (1): 201–211. doi:10.1214/aoms/1177706717. JSTOR 2237306..
40. http://sci.tech-archive.net/Archive/sci.stat.math/2006-02/msg00171.html
41. A MATLAB Toolbox for computing Weighted Correlation Coefficients
42. Nikolić, D; Muresan, RC; Feng, W; Singer, W (2012). "Scaled correlation analysis: a better way to compute a cross-correlogram" (PDF). European Journal of Neuroscience: 1–21. doi:10.1111/j.1460-9568.2011.07987.x.
43. Fulekar (Ed.), M.H. (2009) Bioinformatics: Applications in Life and Environmental Sciences, Springer (pp. 110) ISBN 1-4020-8879-5
44. Jammalamadaka, S. Rao; SenGupta, A. (2001). Topics in circular statistics. New Jersey: World Scientific. p. 176. ISBN 981-02-3778-2. Retrieved 2016-09-21.
45. Moriya, N. (2008). Fengshan Yang, ed. "Noise-Related Multivariate Optimal Joint-Analysis in Longitudinal Stochastic Processes". Nova Science Publishers, Inc.: 223–260. ISBN 978-1-60021-976-4.
46. Garren, Steven T (15 June 1998). "Maximum likelihood estimation of the correlation coefficient in a bivariate normal model with missing data". Statistics & Probability Letters. Elsevier. 38 (3): 281–288. doi:10.1016/S0167-7152(98)00035-2. Retrieved 5 December 2015.
47. Cox, D.R., Hinkley, D.V. (1974) Theoretical Statistics, Chapman & Hall (Appendix 3) ISBN 0-412-12420-3

CHAPTER 6

Regression Analysis

Chapter Contents

- Introduction
- History
- Necessary Number of Independent Measurements
- Linear Regression
- Nonlinear Regression
- Local Regression
- Robust Regression
- Segmented Regression
- Stepwise Regression

Regression is a statistical measure used in finance, investing and other disciplines that attempts to determine the strength of the relationship between one dependent variable (usually denoted by Y) and a series of other changing variables (known as independent variables). Regression helps investment and financial managers to value assets and understand the relationships between variables, such as commodity prices and the stocks of businesses dealing in those commodities. Regression analysis is used to model the relationship between a response variable and one or more predictor variables.

INTRODUCTION

In statistical modeling, regression analysis is a statistical process for estimating the relationships among variables. It includes many techniques for modeling and analyzing several variables, when the focus is on the relationship between a dependent variable and one or more independent variables (or 'predictors'). More specifically, regression analysis helps one understand how the typical value of the dependent variable (or 'criterion variable') changes when any one of the independent variables is varied, while the other independent variables are held fixed. Most commonly, regression analysis estimates the conditional expectation of the dependent variable given the independent variables – that is, the average value of the dependent variable when the independent variables are fixed. Less commonly, the focus is on a quantile, or other location parameter of the conditional distribution of the dependent variable given the independent variables. In all cases, the estimation target is a function of the independent variables called the regression function. In regression analysis, it is also of interest to characterize the variation of the dependent variable around the regression function which can be described by a probability distribution. A related but distinct approach is necessary condition analysis (NCA), which estimates the maximum (rather than average) value of the dependent variable for a given value of the independent variable (ceiling line rather than central line) in order to identify what value of the independent variable is necessary but not sufficient for a given value of the dependent variable.

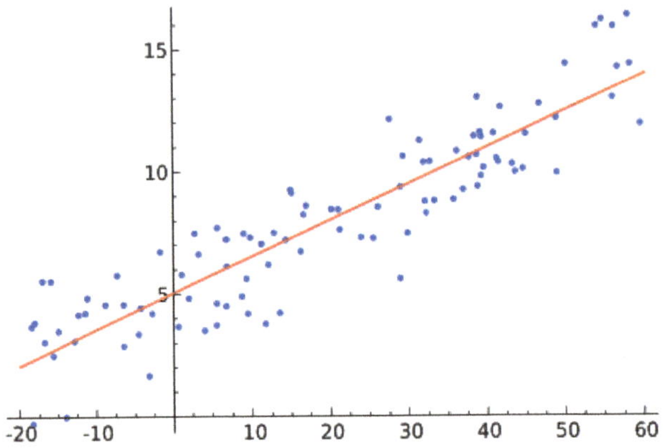

Regression analysis

Regression analysis is widely used for prediction and forecasting, where its use has substantial overlap with the field of machine learning. Regression analysis is also used to understand which among the independent variables are related to the dependent variable, and to explore the forms of these relationships. In restricted circumstances, regression analysis can be used to infer causal relationships between the independent and dependent variables. However this can lead to illusions or false relationships, so caution is advisable; for example, correlation does not imply causation.

Many techniques for carrying out regression analysis have been developed. Familiar methods such as linear regression and ordinary least squares regression are parametric, in that the regression function is defined in terms of a finite number of unknown parameters that are estimated from the data. Nonparametric regression refers to techniques that allow the regression function to lie in a specified set of functions, which may be infinite-dimensional.

The performance of regression analysis methods in practice depends on the form of the data generating process, and how it relates to the regression approach being used. Since the true form of the data-generating process is generally not known, regression analysis often depends to some extent on making assumptions about this process. These assumptions are sometimes testable if a sufficient quantity of data is available. Regression models for prediction are often useful even when the assumptions are moderately violated, although they may not perform optimally. However, in many applications, especially with small effects or questions of causality based on observational data, regression methods can give misleading results.

In a narrower sense, regression may refer specifically to the estimation of continuous response variables, as opposed to the discrete response variables used in classification. The case of a continuous output variable may be more specifically referred to as metric regression to distinguish it from related problems.

HISTORY

The earliest form of regression was the method of least squares, which was published by Legendre in 1805, and by Gauss in 1809. Legendre and Gauss both applied the method to the problem of determining, from astronomical observations, the orbits of bodies about the Sun (mostly comets, but also later the then newly discovered minor planets). Gauss published a further development of the theory of least squares in 1821, including a version of the Gauss–Markov theorem.

The term "regression" was coined by Francis Galton in the nineteenth century to describe a biological phenomenon. The phenomenon was that the heights of descendants of tall ancestors tend to regress down towards a normal average (a phenomenon also known as regression toward the mean). For Galton, regression had only this biological meaning, but his work was later extended by Udny Yule and Karl Pearson to a more general statistical context. In the work of Yule and Pearson, the joint distribution of the response and explanatory variables is assumed to be Gaussian. This assumption was weakened by R.A. Fisher in his works of 1922 and 1925. Fisher assumed that the conditional distribution of the response variable is Gaussian, but the joint distribution need not be. In this respect, Fisher's assumption is closer to Gauss's formulation of 1821.

In the 1950s and 1960s, economists used electromechanical desk calculators to calculate regressions. Before 1970, it sometimes took up to 24 hours to receive the result from one regression.

Regression methods continue to be an area of active research. In recent decades, new methods have been developed for robust regression, regression involving correlated responses such as time series and growth curves, regression in which the predictor (independent variable) or response variables are curves, images, graphs, or other complex data objects, regression methods accommodating various types of missing data, nonparametric regression, Bayesian methods for regression, regression in which the predictor variables are measured with error, regression with more predictor variables than observations, and causal inference with regression.

Regression models

Regression models involve the following variables:

- The unknown parameters, denoted as $\boldsymbol{\beta}$, which may represent a scalar or a vector.
- The independent variables, X.

- The dependent variable, Y.

In various fields of application, different terminologies are used in place of dependent and independent variables.

A regression model relates Y to a function of $\mathbf{X}$ and $\boldsymbol{\beta}$.

$Y \approx f(X,\beta)$

The approximation is usually formalized as $E(Y \mid \mathbf{X}) = f(\mathbf{X}, \boldsymbol{\beta})$. To carry out regression analysis, the form of the function f must be specified. Sometimes the form of this function is based on knowledge about the relationship between Y and $\mathbf{X}$ that does not rely on the data. If no such knowledge is available, a flexible or convenient form for f is chosen.

Assume now that the vector of unknown parameters $\boldsymbol{\beta}$ is of length k. In order to perform a regression analysis the user must provide information about the dependent variable Y:

- If N data points of the form $(Y, \mathbf{X})$ are observed, where $N < k$, most classical approaches to regression analysis cannot be performed: since the system of equations defining the regression model is underdetermined, there are not enough data to recover $\boldsymbol{\beta}$.
- If exactly $N = k$ data points are observed, and the function f is linear, the equations $Y = f(\mathbf{X}, \boldsymbol{\beta})$ can be solved exactly rather than approximately. This reduces to solving a set of N equations with N unknowns (the elements of $\boldsymbol{\beta}$), which has a unique solution as long as the $\mathbf{X}$ are linearly independent. If f is nonlinear, a solution may not exist, or many solutions may exist.
- The most common situation is where $N > k$ data points are observed. In this case, there is enough information in the data to estimate a unique value for $\boldsymbol{\beta}$ that best fits the data in some sense, and the regression model when applied to the data can be viewed as an over determined system in $\boldsymbol{\beta}$.

In the last case, the regression analysis provides the tools for:

- Finding a solution for unknown parameters $\boldsymbol{\beta}$ that will, for example, minimize the distance between the measured and predicted values of the dependent variable Y (also known as method of least squares).
- Under certain statistical assumptions, the regression analysis uses the surplus of information to provide statistical information about the unknown parameters $\boldsymbol{\beta}$ and predicted values of the dependent variable Y.

NECESSARY NUMBER OF INDEPENDENT MEASUREMENTS

Consider a regression model which has three unknown parameters, β_0, β_1, and β_2. Suppose an experimenter performs 10 measurements all at exactly the same value of independent variable vector $\mathbf{X}$ (which contains the independent variables X_1, X_2, and X_3). In this case, regression analysis fails to give a unique set of estimated values for the three unknown parameters; the experimenter did not provide enough information. The best one can do is to estimate the average value and the standard deviation of the dependent variable Y.

Similarly, measuring at two different values of **X** would give enough data for a regression with two unknowns, but not for three or more unknowns.

If the experimenter had performed measurements at three different values of the independent variable vector **X**, then regression analysis would provide a unique set of estimates for the three unknown parameters in $\boldsymbol{\beta}$.

In the case of general linear regression, the above statement is equivalent to the requirement that the matrix $\mathbf{X}^T\mathbf{X}$ is invertible.

Statistical Assumptions

When the number of measurements, N, is larger than the number of unknown parameters, k, and the measurement errors ε_i are normally distributed then *the excess of information* contained in $(N - k)$ measurements is used to make statistical predictions about the unknown parameters. This excess of information is referred to as the degrees of freedom of the regression.

Underlying Assumptions

Classical assumptions for regression analysis include:

- The sample is representative of the population for the inference prediction.
- The error is a random variable with a mean of zero conditional on the explanatory variables.
- The independent variables are measured with no error. (Note: If this is not so, modeling may be done instead using errors-in-variables model techniques).
- The independent variables (predictors) are linearly independent, i.e. it is not possible to express any predictor as a linear combination of the others.
- The errors are uncorrelated, that is, the variance–covariance matrix of the errors is diagonal and each non-zero element is the variance of the error.
- The variance of the error is constant across observations (homoscedasticity). If not, weighted least squares or other methods might instead be used.

These are sufficient conditions for the least-squares estimator to possess desirable properties; in particular, these assumptions imply that the parameter estimates will be unbiased, consistent, and efficient in the class of linear unbiased estimators. It is important to note that actual data rarely satisfies the assumptions. That is, the method is used even though the assumptions are not true. Variation from the assumptions can sometimes be used as a measure of how far the model is from being useful. Many of these assumptions may be relaxed in more advanced treatments. Reports of statistical analyses usually include analyses of tests on the sample data and methodology for the fit and usefulness of the model.

Assumptions include the geometrical support of the variables. Independent and dependent variables often refer to values measured at point locations. There may be spatial trends and spatial autocorrelation in the variables that violate statistical assumptions of regression. Geographic weighted regression is one technique to deal with such data. Also, variables may include values aggregated by areas. With aggregated data the modifiable areal unit problem can cause extreme variation in regression parameters. When analyzing data aggregated by political boundaries, postal codes or census areas results may be very distinct with a different choice of units.

LINEAR REGRESSION

In linear regression, the model specification is that the dependent variable, y_i is a linear combination of the *parameters* (but need not be linear in the *independent variables*). For example, in simple linear regression for modeling n data points there is one independent variable: x_i, and two parameters, β_0 and β_1

straight line:

$$y_i = \beta_0 + \beta_1 x_i + \varepsilon_i, \quad i = 1, \dots, n.$$

In multiple linear regression, there are several independent variables or functions of independent variables.

Adding a term in x_i^2 to the preceding regression gives:

parabola:

$$y_i = \beta_0 + \beta_1 x_i + \beta_2 x_i^2 + \varepsilon_i,\ i = 1, \dots, n.$$

This is still linear regression; although the expression on the right hand side is quadratic in the independent variable x_i, it is linear in the parameters β_0, β_1 and β_2.

In both cases, ε_i is an error term and the subscript i indexes a particular observation.

Returning our attention to the straight line case: Given a random sample from the population, we estimate the population parameters and obtain the sample linear regression model:

$$\widehat{y_i} = \widehat{\beta}_0 + \widehat{\beta}_1 x_i.$$

The residual, $e_i = y_i - \hat{y}_i$, is the difference between the value of the dependent variable predicted by the model, $\widehat{y_i}$, and the true value of the dependent variable, y_i. One method of estimation is ordinary least squares. This method obtains parameter estimates that minimize the sum of squared residuals, SSE, also sometimes denoted RSS:

$$SSE = \sum_{i=1}^{n} e_i^2.$$

Minimization of this function results in a set of normal equations, a set of simultaneous linear equations in the parameters, which are solved to yield the parameter estimators, $\widehat{\beta}_0, \widehat{\beta}_1$.

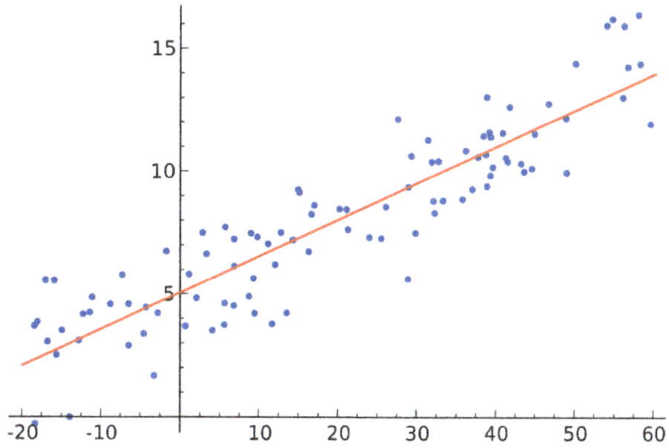

Illustration of linear regression on a data set.

In the case of simple regression, the formulas for the least squares estimates are

$$\widehat{\beta_1} = \frac{\sum(x_i - \bar{x})(y_i - \bar{y})}{\sum(x_i - \bar{x})^2} \text{ and } \widehat{\beta_0} = \bar{y} - \widehat{\beta_1}\bar{x}$$

where $\bar{x}$ is the mean (average) of the x values and $\bar{y}$ is the mean of the y values.

Under the assumption that the population error term has a constant variance, the estimate of that variance is given by:

$$\hat{\sigma}_\varepsilon^2 = \frac{SSE}{n-2}.$$

This is called the mean square error (MSE) of the regression. The denominator is the sample size reduced by the number of model parameters estimated from the same data, (n-p) for p regressors or (n-p-1) if an intercept is used. In this case, p=1 so the denominator is n-2.

The standard errors of the parameter estimates are given by

$$\hat{\sigma}_{\beta_0} = \hat{\sigma}_\varepsilon \sqrt{\frac{1}{n} + \frac{\bar{x}^2}{\sum(x_i - \bar{x})^2}}$$

$$\hat{\sigma}_{\beta_1} = \hat{\sigma}_\varepsilon \sqrt{\frac{1}{\sum(x_i - \bar{x})^2}}.$$

Under the further assumption that the population error term is normally distributed, the researcher can use these estimated standard errors to create confidence intervals and conduct hypothesis tests about the population parameters.

General Linear Model

In the more general multiple regression model, there are p independent variables:

$$y_i = \beta_1 x_{i1} + \beta_2 x_{i2} + \cdots + \beta_p x_{ip} + \varepsilon_i,$$

where x_{ij} is the ith observation on the jth independent variable. If the first independent variable takes the value 1 for all i, $x_{i1} = 1$, then β_1 is called the regression intercept.

The least squares parameter estimates are obtained from p normal equations. The residual can be written as

$$\sum_{i=1}^{n}\sum_{k=1}^{p} X_{ij}X_{ik}\hat{\beta}_k = \sum_{i=1}^{n} X_{ij}y_i,\ j = 1,\ldots,p.$$

In matrix notation, the normal equations are written as

$$(\mathbf{X}^\top\mathbf{X})\hat{\beta} = \mathbf{X}^\top\mathbf{Y},$$

where the ij element of X is x_{ij}, the i element of the column vector Y is y_i, and the j element of $\hat{\beta}$ is $\hat{\beta}_j$. Thus X is $n{\times}p$, Y is $n{\times}1$, and $\hat{\beta}$ is $p{\times}1$. The solution is

$$\hat{\beta} = (\mathbf{X}^\top\mathbf{X})^{-1}\mathbf{X}^\top\mathbf{Y}.$$

Diagnostics

Once a regression model has been constructed, it may be important to confirm the goodness of fit of the model and the statistical significance of the estimated parameters. Commonly used checks of goodness of fit include the R-squared, analyses of the pattern of residuals and hypothesis testing. Statistical significance can be checked by an F-test of the overall fit, followed by t-tests of individual parameters.

Interpretations of these diagnostic tests rest heavily on the model assumptions. Although examination of the residuals can be used to invalidate a model, the results of a t-test or F-testare sometimes more difficult to interpret if the model's assumptions are violated. For example, if the error term does not have a normal distribution, in small samples the estimated parameters will not follow normal distributions and complicate inference. With relatively large samples, however, a central limit theorem can be invoked such that hypothesis testing may proceed using asymptotic approximations.

"Limited Dependent" Variables

The phrase "limited dependent" is used in econometric statistics for categorical and constrained variables.

The response variable may be non-continuous ("limited" to lie on some subset of the real line). For binary (zero or one) variables, if analysis proceeds with least-squares linear regression, the model is called the linear probability model. Nonlinear models for binary dependent variables include the probit and logit model. The multivariate probit model is a standard method of estimating a joint relationship between several binary dependent variables and some independent variables. For categorical variables with more than two values there is the multinomial logit. For ordinal variables with more than two values, there are the ordered logit and ordered probit models. Censored regression models may be used when the dependent variable is only sometimes observed, and Heckman correction type models may be used when the sample is not randomly selected from the population of interest. An alternative to such procedures is linear regression based on polychoric correlation (or polyserial correlations) between the categorical variables. Such procedures differ in the assumptions made about the distribution of the variables in the population. If the variable is positive with low values and represents the repetition of the occurrence of an event, then count models like the Poisson regression or the negative binomial model may be used instead.

Interpolation and Extrapolation

Regression models predict a value of the Y variable given known values of the X variables. Prediction *within* the range of values in the dataset used for model-fitting is known informally as interpolation. Prediction *outside* this range of the data is known as extrapolation. Performing extrapolation relies strongly on the regression assumptions. The further the extrapolation goes outside the data, the more room there is for the model to fail due to differences between the assumptions and the sample data or the true values.

It is generally advised that when performing extrapolation, one should accompany the estimated value of the dependent variable with a prediction interval that represents the uncertainty. Such intervals tend to expand rapidly as the values of the independent variable(s) moved outside the range covered by the observed data.

For such reasons and others, some tend to say that it might be unwise to undertake extrapolation.

However, this does not cover the full set of modelling errors that may be being made: in particular, the assumption of a particular form for the relation between Y and X. A properly conducted regression analysis will include an assessment of how well the assumed form is matched by the observed data, but it can only do so within the range of values of the independent variables actually available. This means that any extrapolation is particularly reliant on the assumptions being made about the structural form of the regression relationship. Best-practice advice here is that a linear-in-variables and linear-in-parameters relationship should not be chosen simply for computational convenience, but that all available knowledge should be deployed in constructing a regression model. If this knowledge includes the fact that the dependent variable cannot go outside a certain range of values, this can be made use of in selecting the model – even if the observed dataset has no values particularly near such bounds. The implications of this step of choosing an appropriate functional form for the regression can be great when extrapolation is considered. At a minimum, it can ensure that any extrapolation arising from a fitted model is "realistic" (or in accord with what is known).

NONLINEAR REGRESSION

When the model function is not linear in the parameters, the sum of squares must be minimized by an iterative procedure. This introduces many complications which are summarized in Differences between linear and non-linear least squares

Power and Sample Size Calculations

There are no generally agreed methods for relating the number of observations versus the number of independent variables in the model. One rule of thumb suggested by Good and Hardin is $N = m^n$, where N is the sample size, n is the number of independent variables and m is the number of observations needed to reach the desired precision if the model had only one independent variable. For example, a researcher is building a linear regression model using a dataset that contains 1000 patients N. If the researcher decides that five observations are needed to precisely define a straight line m, then the maximum number of independent variables the model can support is 4, because

$$\frac{\log 1000}{\log 5} = 4.29$$

Other Methods

Although the parameters of a regression model are usually estimated using the method of least squares, other methods which have been used include:

- Bayesian methods, e.g. Bayesian linear regression
- Percentage regression, for situations where reducing *percentage* errors is deemed more appropriate.
- Least absolute deviations, which is more robust in the presence of outliers, leading to quantile regression
- Nonparametric regression, requires a large number of observations and is computationally intensive
- Distance metric learning, which is learned by the search of a meaningful distance metric in a given input space.

Software

All major statistical software packages perform least squares regression analysis and inference. Simple linear regression and multiple regression using least squares can be done in some spreadsheet applications and on some calculators. While many statistical software packages can perform various types of nonparametric and robust regression, these methods are less standardized; different software packages implement different methods, and a method with a given name may be implemented differently in different packages. Specialized regression software has been developed for use in fields such as survey analysis and neuroimaging.

LOCAL REGRESSION

LOESS and LOWESS (locally weighted scatterplot smoothing) are two strongly related non-parametric regressionmethods that combine multiple regression models in a *k*-nearest-neighbor-based meta-model. "LOESS" is a later generalization of LOWESS; although it is not a true initialism, it may be understood as standing for "LOcal regrESSion".

LOESS and LOWESS thus build on "classical" methods, such as linear and nonlinear least squares regression. They address situations in which the classical procedures do not perform well or cannot be effectively applied without undue labor. LOESS combines much of the simplicity of linear least squares regression with the flexibility of nonlinear regression. It does this by fitting simple models to localized subsets of the data to build up a function that describes the deterministic part of the variation in the data, point by point. In fact, one of the chief attractions of this method is that the data analyst is not required to specify a global function of any form to fit a model to the data, only to fit segments of the data.

The trade-off for these features is increased computation. Because it is so computationally intensive, LOESS would have been practically impossible to use in the era when least squares regression was being developed. Most other modern methods for process modeling are similar to LOESS in this respect. These methods have been consciously designed to use our current computational ability to the fullest possible advantage to achieve goals not easily achieved by traditional approaches.

A smooth curve through a set of data points obtained with this statistical technique is called a Loess Curve, particularly when each smoothed value is given by a weighted quadratic least squares regression over the span of values of the y-axis scattergram criterion variable. When each smoothed value is given by a weighted linear least squares regression over the span, this is known as a Lowess curve; however, some authorities treat Lowess and Loess as synonyms.

Definition of a LOESS Model

LOESS, originally proposed by Cleveland (1979) and further developed by Cleveland and Devlin (1988), specifically denotes a method that is also known as locally weighted polynomial regression. At each point in the range of the data set a low-degree polynomial is fitted to a subset of the data, with explanatory variable values near the point whose response is being estimated. The polynomial is fitted using weighted least squares, giving more weight to points near the point whose response is being estimated and less weight to points further away. The value of the regression function for the point is then obtained by evaluating the local polynomial using the explanatory variable values for that data point. The LOESS fit is complete after regression function values have been computed for each of the n data points. Many of the details of this method, such as the degree of the polynomial model and the weights, are flexible. The range of choices for each part of the method and typical defaults are briefly discussed next.

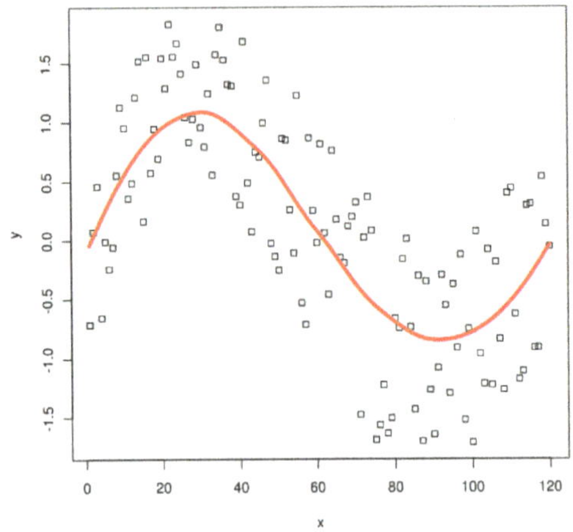

LOESS curve fitted to a population sampled from a sine wave with uniform noise added. The LOESS curve approximates the original sine wave

Localized Subsets of Data

The **subsets** of data used for each weighted least squares fit in LOESS are determined by a nearest neighbors algorithm. A user-specified input to the procedure called the "bandwidth" or "smoothing parameter" determines how much of the data is used to fit each local polynomial. The smoothing parameter, *a* is a number between $(\lambda + 1)/n$ and 1, with λ denoting the degree of the local polynomial. The value of a is the proportion of data used in each fit. The subset of data used in each weighted least squares fit comprises the n a points (rounded to the next largest integer) whose explanatory variables' values are closest to the point at which the response is being estimated.

a is called the smoothing parameter because it controls the flexibility of the LOESS regression function. Large values of *a* produce the smoothest functions that wiggle the least in response to fluctuations in the data. The smaller *a* is, the closer the regression function will conform to the data. Using too small a value of the smoothing parameter is not desirable, however, since the regression function will eventually start to capture the random error in the data. Useful values of the smoothing parameter typically lie in the range 0.25 to 0.5 for most LOESS applications.

Degree of Local Polynomials

The local polynomials fit to each subset of the data are almost always of first or second degree; that is, either locally linear (in the straight line sense) or locally quadratic. Using a zero degree polynomial turns LOESS into a weighted moving average. Such a simple local model might work well for some situations, but may not always approximate the underlying function well enough. Higher-degree polynomials would work in theory, but yield models that are not really in the spirit of LOESS. LOESS is based on the ideas that any function can be well approximated in a small neighborhood by a low-order polynomial and that simple models can be fit to data easily. High-degree polynomials would tend to overfit the data in each subset and are numerically unstable, making accurate computations difficult.

Weight Function

As mentioned above, the weight function gives the most weight to the data points nearest the point of estimation and the least weight to the data points that are furthest away. The use of the weights is based on the idea that points near each other in the explanatory variable space are more likely to be related to each other in a simple way than points that are further apart. Following this logic, points that are likely to follow the local model best influence the local model parameter estimates the most. Points that are less likely to actually conform to the local model have less influence on the local model parameter estimates.

The traditional weight function used for LOESS is the tri-cube weight function,

$$w(x) = (1 - |x|^3)^3 \, \mathrm{I}[|x| < 1]$$

However, any other weight function that satisfies the properties listed in Cleveland (1979) could also be used. The weight for a specific point in any localized subset of data is obtained by evaluating the weight function at the distance between that point and the point of estimation, after scaling the distance so that the maximum absolute distance over all of the points in the subset of data is exactly one.

Advantages of LOESS

As discussed above, the biggest advantage LOESS has over many other methods is the fact that it does not require the specification of a function to fit a model to all of the data in the sample. Instead the analyst only has to provide a smoothing parameter value and the degree of the local polynomial. In addition, LOESS is very flexible, making it ideal for modeling complex processes for which no theoretical models exist. These two advantages, combined with the simplicity of the method, make LOESS one of the most attractive of the modern regression methods for applications that fit the general framework of least squares regression but which have a complex deterministic structure.

Although it is less obvious than for some of the other methods related to linear least squares regression, LOESS also accrues most of the benefits typically shared by those procedures. The most important of those is the theory for computing uncertainties for prediction and calibration. Many other tests and procedures used for validation of least squares models can also be extended to LOESS models.

Disadvantages of LOESS

LOESS makes less efficient use of data than other least squares methods. It requires fairly large, densely sampled data sets in order to produce good models. This is because LOESS relies on the local data structure when performing the local fitting. Thus, LOESS provides less complex data analysis in exchange for greater experimental costs.

Another disadvantage of LOESS is the fact that it does not produce a regression function that is easily represented by a mathematical formula. This can make it difficult to transfer the results of an analysis to other people. In order to transfer the regression function to another person, they would need the data set and software for LOESS calculations. In nonlinear regression, on the other hand, it is only necessary to write down a functional form in order to provide estimates of the unknown parameters and the estimated uncertainty. Depending on the application, this could be either a major or a minor drawback to using LOESS. In particular, the simple form of LOESS cannot be used for mechanistic modelling where fitted parameters specify particular physical properties of a system.

Finally, as discussed above, LOESS is a computationally intensive method (with the exception of evenly spaced data, where the regression can then be phrased as a Non-CausalFinite impulse response filter). LOESS is also prone to the effects of outliers in the data set, like other least squares methods. There is an iterative, robust version of LOESS [Cleveland (1979)] that can be used to reduce LOESS' sensitivity to outliers, but too many extreme outliers can still overcome even the robust method.

ROBUST REGRESSION

In robust statistics, robust regression is a form of regression analysis designed to circumvent some limitations of traditional parametric and non-parametric methods. Regression analysis seeks to find the relationship between one or more independent variables and a dependent variable. Certain widely used methods of regression, such as ordinary least squares, have favorable properties if their underlying assumptions are true, but can give misleading results if those assumptions are not true; thus ordinary least squares is said to be not robust to violations of its assumptions. Robust regression methods are designed to be not overly affected by violations of assumptions by the underlying data-generating process.

In particular, least squares estimates for regression models are highly sensitive to (not robust against) outliers. While there is no precise definition of an outlier, outliers are observations which do not follow the pattern of the other observations. This is not normally a problem if the outlier is simply an extreme observation drawn from the tail of a normal distribution, but if the outlier results from non-normal measurement error or some other violation of standard ordinary least squares assumptions, then it compromises the validity of the regression results if a non-robust regression technique is used.

Applications

Heteroscedastic Errors

One instance in which robust estimation should be considered is when there is a strong suspicion of heteroscedasticity. In thehomoscedastic model, it is assumed that the variance of the error term is constant for all values of x. Heteroscedasticity allows the variance to be dependent on x, which is more accurate for many real scenarios. For example, the variance of expenditure is often larger for individuals with higher income than for individuals with lower incomes. Software packages usually default to a homoscedastic model, even though such a model may be less accurate than a heteroscedastic model. One simple approach (Tofallis, 2008) is to apply least squares to percentage errors as this reduces the influence of the larger values of the dependent variable compared to ordinary least squares.

Presence of Outliers

Another common situation in which robust estimation is used occurs when the data contain outliers. In the presence of outliers that do not come from the same data-generating process as the rest of the data, least squares estimation is inefficient and can be biased. Because the least squares predictions are dragged towards the outliers, and because the variance of the estimates is artificially inflated, the result is that outliers can be masked. (In many situations, including some areas of geostatistics and medical statistics, it is precisely the outliers that are of interest.)

Although it is sometimes claimed that least squares (or classical statistical methods in general) are robust, they are only robust in the sense that the type I error rate does not increase under violations of the model. In fact, the type I error rate tends to be lower than the nominal level when outliers are present, and there is often a dramatic increase in the type II error rate. The reduction of the type I error rate has been labelled as the *conservatism* of classical methods.

History and Unpopularity of Robust Regression

Despite their superior performance over least squares estimation in many situations, robust methods for regression are still not widely used. Several reasons may help explain their unpopularity (Hampel et al. 1986, 2005). One possible reason is that there are several competing methods and the field got off to many false starts. Also, computation of robust estimates is much more computationally intensive than least squares estimation; in recent years however, this objection has become less relevant as computing power has increased greatly. Another reason may be that some popular statistical software packages failed to implement the methods (Stromberg, 2004). The belief of many statisticians that classical methods are robust may be another reason.

Although uptake of robust methods has been slow, modern mainstream statistics text books often include discussion of these methods (for example, the books by Seber and Lee, and by Faraway; for a good general description of how the various robust regression methods developed from one another see Andersen's book). Also, modern statistical software packages such as R, Statsmodels, Stata and S-PLUS include considerable functionality for robust estimation (see, for example, the books by Venables and Ripley, and by Maronna et al.).

Methods for Robust Regression

Least Squares Alternatives

The simplest methods of estimating parameters in a regression model that are less sensitive to outliers than the least squares estimates, is to use least absolute deviations. Even then, gross outliers can still have a considerable impact on the model, motivating research into even more robust approaches.

In 1973, Huber introduced M-estimation for regression. The M in M-estimation stands for "maximum likelihood type". The method is robust to outliers in the response variable, but turned out not to be resistant to outliers in the explanatory variables (leverage points). In fact, when there are outliers in the explanatory variables, the method has no advantage over least squares.

In the 1980s, several alternatives to M-estimation were proposed as attempts to overcome the lack of resistance. See the book by Rousseeuw and Leroy for a very practical review. Least trimmed squares (LTS) is a viable alternative and is currently (2007) the preferred choice of Rousseeuw and Ryan (1997, 2008). The Theil–Sen estimator has a lower breakdown point than LTS but is statistically efficient and popular. Another proposed solution was S-estimation. This method finds a line (plane or hyperplane) that minimizes a robust estimate of the scale (from which the method gets the S in its name) of the residuals. This method is highly resistant to leverage points, and is robust to outliers in the response. However, this method was also found to be inefficient.

MM-estimation attempts to retain the robustness and resistance of S-estimation, whilst gaining the efficiency of M-estimation. The method proceeds by finding a highly robust and resistant S-estimate that minimizes an M-estimate of the scale of the residuals (the first M in the method's name). The estimated scale is then held constant whilst a close-by M-estimate of the parameters is located (the second M).

Parametric Alternatives

Another approach to robust estimation of regression models is to replace the normal distribution with a heavy-tailed distribution. A *t*-distribution with between 4 and 6 degrees of freedom has been reported to be a good choice in various practical situations. Bayesian robust regression, being fully parametric, relies heavily on such distributions.

Under the assumption of *t*-distributed residuals, the distribution is a location-scale family. That is, $x \leftarrow (x-\mu)/\sigma$. The degrees of freedom of the *t*-distribution is sometimes called the *kurtosis parameter*. Lange, Little and Taylor (1989) discuss this model in some depth from a non-Bayesian point of view. A Bayesian account appears in Gelman et al. (2003).

An alternative parametric approach is to assume that the residuals follow a mixture of normal distributions; in particular, a *contaminated* normal distribution in which the majority of observations are from a specified normal distribution, but a small proportion are from a normal distribution with much higher variance. That is, residuals have probability $1-\varepsilon$ of coming from a normal distribution with variance σ^2, where ε is small, and probability ε of coming from a normal distribution with variance $c\sigma^2$ for some $c > 1$.

$$e_i \sim (1-\varepsilon)N(0,\sigma^2) + \varepsilon N(0,c\sigma^2).$$

Typically, $\varepsilon < 0.1$. This is sometimes called the ε-contamination model.

Parametric approaches have the advantage that likelihood theory provides an 'off the shelf' approach to inference (although for mixture models such as the ε-contamination model, the usual regularity conditions might not apply), and it is possible to build simulation models from the fit. However, such parametric models still assume that the underlying model is literally true. As such, they do not account for skewed residual distributions or finite observation precisions.

Unit Weights

Another robust method is the use of unit weights (Wainer & Thissen, 1976), a method that can be applied when there are multiple predictors of a single outcome. Ernest Burgess(1928) used unit weights to predict success on parole. He scored 21 positive factors as present (e.g., "no prior arrest" = 1) or absent ("prior arrest" = 0), then summed to yield a predictor score, which was shown to be a useful predictor of parole success. Samuel S. Wilks (1938) showed that nearly all sets of regression weights sum to composites that are very highly correlated with one another, including unit weights, a result referred to as Wilk's theorem (Ree, Carretta, & Earles, 1998). Robyn Dawes (1979) examined decision making in applied settings, showing that simple models with unit weights often outperformed human experts. Bobko, Roth, and Buster (2007) reviewed the literature on unit weights, and they concluded that decades of empirical studies show that unit weights perform similar to ordinary regression weights on cross validation.

Example: BUPA Liver Data

The BUPA liver data have been studied by various authors, including Breiman (2001). The data can be found via the classic data sets page and there is some discussion in the article on the Box-Cox transformation. A plot of the logs of ALT versus the logs of γGT appears below. The two regression lines are those estimated by ordinary least squares (OLS) and by robust MM-estimation. The analysis was performed in R using software made available by Venables and Ripley (2002).

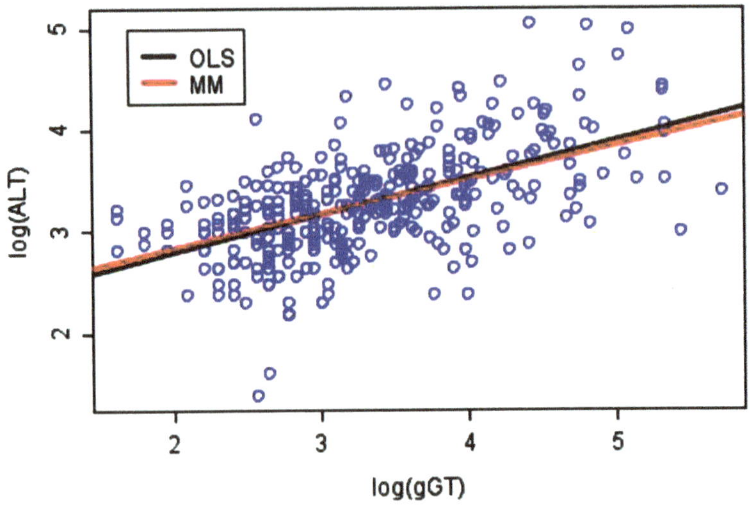

The two regression lines appear to be very similar (and this is not unusual in a data set of this size). However, the advantage of the robust approach comes to light when the estimates of residual scale are considered. For ordinary least squares, the estimate of scale is 0.420, compared to 0.373 for the robust method. Thus, the relative efficiency of ordinary least squares to MM-estimation in this example is 1.266. This inefficiency leads to loss of power in hypothesis tests, and to unnecessarily wide confidence intervals on estimated parameters.

Outlier Detection

Another consequence of the inefficiency of the ordinary least squares fit is that several outliers are masked. Because the estimate of residual scale is inflated, the scaled residuals are pushed closer to zero than when a more appropriate estimate of scale is used. The plots of the scaled residuals from the two models appear below. The variable on the x-axis is just the observation number as it appeared in the data set. Rousseeuw and Leroy (1986) contains many such plots.

Whilst in one or two dimensions outlier detection using classical methods can be performed manually, with large data sets and in high dimensions the problem of masking can make identification of many outliers impossible. Robust methods automatically detect these observations, offering a serious advantage over classical methods when outliers are present.

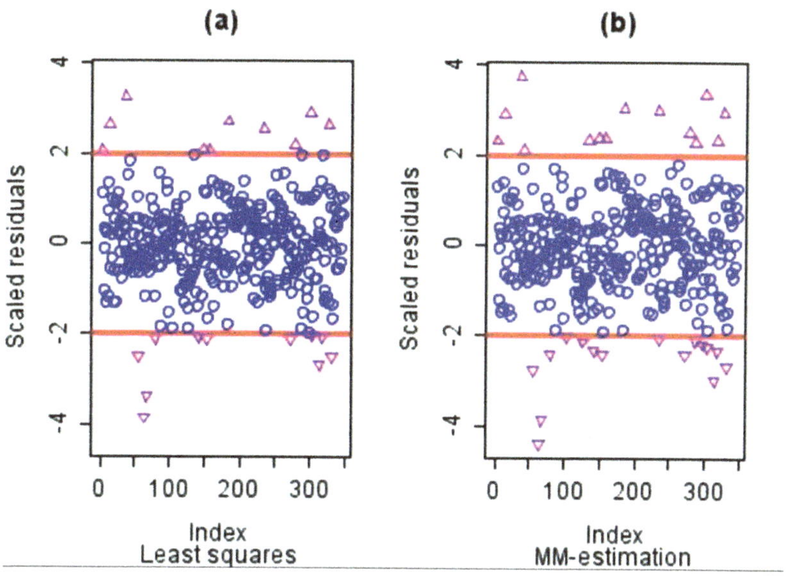

The horizontal reference lines are at 2 and -2 so that any observed scaled residual beyond these boundaries can be considered to be an outlier. Clearly, the least squares method leads to many interesting observations being masked.

SEGMENTED REGRESSION

Segmented regression, also known as piecewise regression or 'broken-stick regression', is a method in regression analysis in which the independent variable is partitioned into intervals and a separate line segment is fit to each interval. Segmented regression analysis can also be performed on multivariate data by partitioning the various independent variables. Segmented regression is useful when the independent variables, clustered into different groups, exhibit different relationships between the variables in these regions. The boundaries between the segments are *breakpoints*.

Segmented linear regression is segmented regression whereby the relations in the intervals are obtained by linear regression.

Segmented Linear Regression, Two Segments

Segmented linear regression with two segments separated by a *breakpoint* can be useful to quantify an abrupt change of the response function (Yr) of a varying influential factor (**x**). The breakpoint can be interpreted as a *critical*, *safe*, or *threshold* value beyond or below which (un)desired effects occur. The breakpoint can be important in decision making

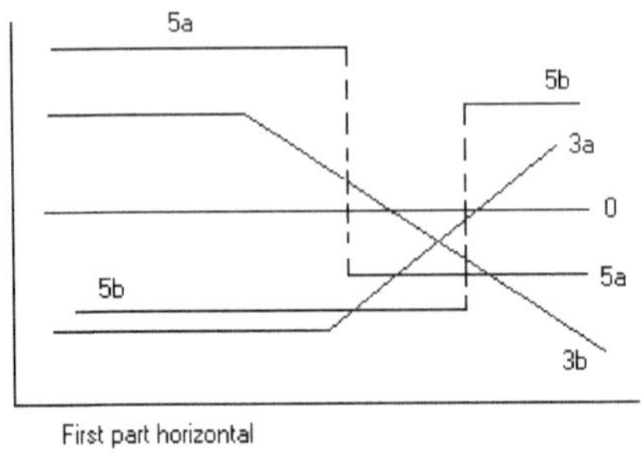

1st limb horizontal

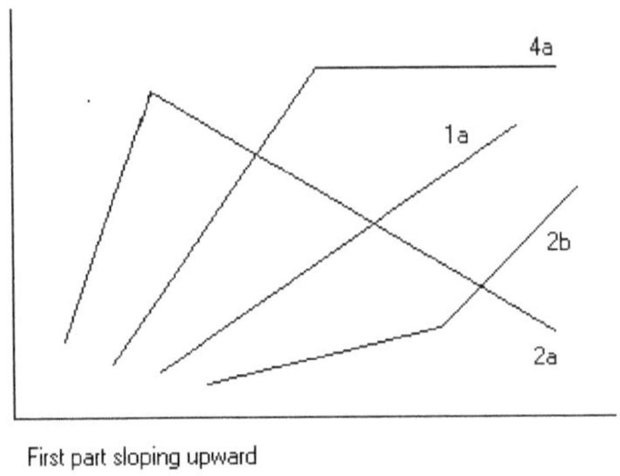

1st limb sloping up

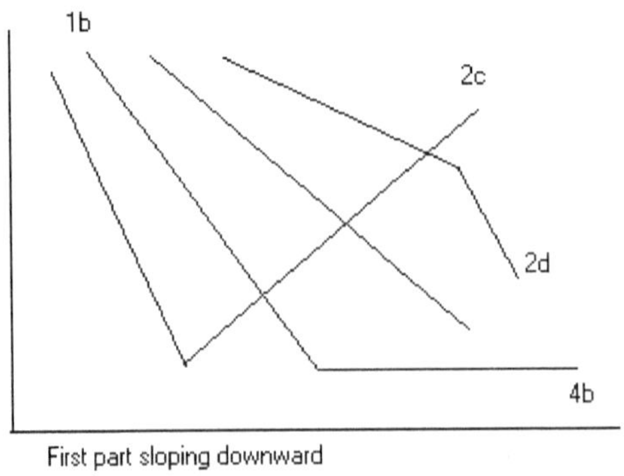

1st limb sloping down

The figures illustrate some of the results and regression types obtainable.

A segmented regression analysis is based on the presence of a set of (**y, x**) data, in which **y** is the dependent variable and **x** the independent variable.

The least squares method applied separately to each segment, by which the two regression lines are made to fit the data set as closely as possible while minimizing the *sum of squares of the differences* (SSD) between observed (**y**) and calculated (Yr) values of the dependent variable, results in the following two equations:

- $Yr = A_1.\mathbf{x} + K_1$ for **x** < BP (breakpoint)
- $Yr = A_2.\mathbf{x} + K_2$ for **x** > BP (breakpoint)

where:

Yr is the expected (predicted) value of **y** for a certain value of **x**;

A_1 and A_2 are regression coefficients (indicating the slope of the line segments);

K_1 and K_2 are *regression constants* (indicating the intercept at the **y**-axis).

The data may show many types or trends, see the figures.

The method also yields two correlation coefficients (R):

$$R_1^2 = 1 - \frac{\sum(y - Y_r)^2}{\sum(y - Y_{a1})^2} \quad \text{for x < BP (breakpoint)}$$

and

$$R_2^2 = 1 - \frac{\sum(y - Y_r)^2}{\sum(y - Y_{a2})^2} \quad \text{for x > BP (breakpoint)}$$

where:

$\sum(y - Y_r)^2$ is the minimized SSD per segment

and

Y_{a1} and Y_{a2} are the average values of y in the respective segments.

In the determination of the most suitable trend, statistical tests must be performed to ensure that this trend is reliable (significant).

When no significant breakpoint can be detected, one must fall back on a regression without breakpoint.

Example

For the blue figure at the right that gives the relation between yield of mustard (Yr = Ym, t/ha) and soil salinity (**x** = Ss, expressed as electric conductivity of the soil solution EC in dS/m) it is found that:

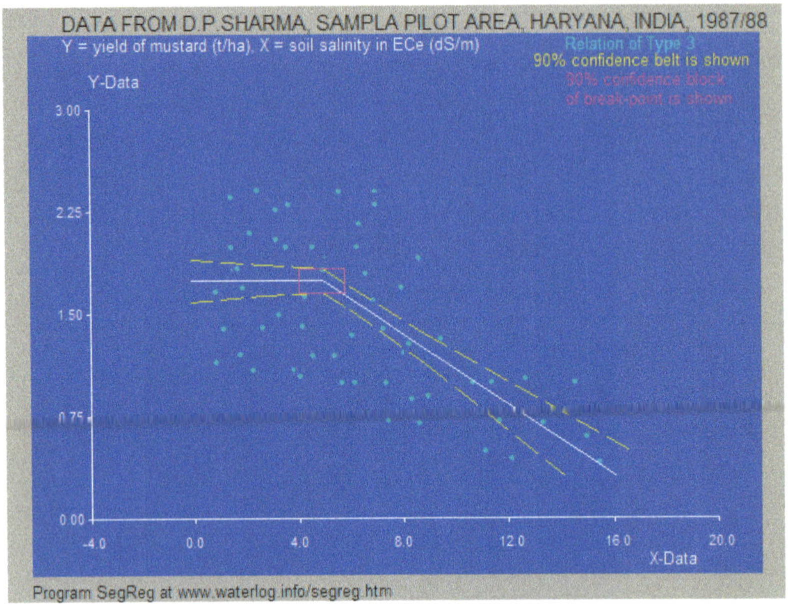

Segmented linear regression, type 3b

BP = 4.93, A_1 = 0, K_1 = 1.74, A_2 = −0.129, K_2 = 2.38, R_1^2 = 0.0035 (insignificant), R_2^2 = 0.395 (significant) and:

- Ym = 1.74 t/ha for Ss < 4.93 (breakpoint)
- Ym = −0.129 Ss + 2.38 t/ha   for Ss > 4.93 (breakpoint)

indicating that soil salinities < 4.93 dS/m are safe and soil salinities > 4.93 dS/m reduce the yield @ 0.129 t/ha per unit increase of soil salinity.

The figure also shows confidence intervals and uncertainty as elaborated hereunder.

Test Procedures

The following *statistical tests* are used to determine the type of trend:

- significance of the breakpoint (BP) by expressing BP as a function of *regression coefficients* A_1 and A_2 and the means Y_1 and Y_2 of the **y**-data and the means X_1 and X_2 of the **x** data (left and right of BP), using the laws of propagation of errors in additions and multiplications to compute the standard error (SE) of BP, and applying Student's t-test
- significance of A_1 and A_2 applying Student's t-distribution and the *standard error* SE of A_1 and A_2
- significance of the difference of A_1 and A_2 applying Student's t-distribution using the SE of their difference.
- significance of the difference of Y_1 and Y_2 applying Student's t-distribution using the SE of their difference.

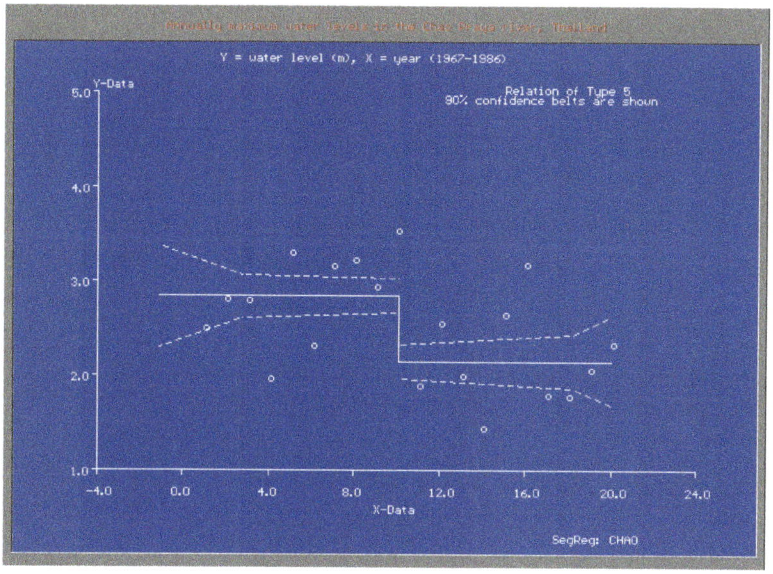

Example time series, type 5

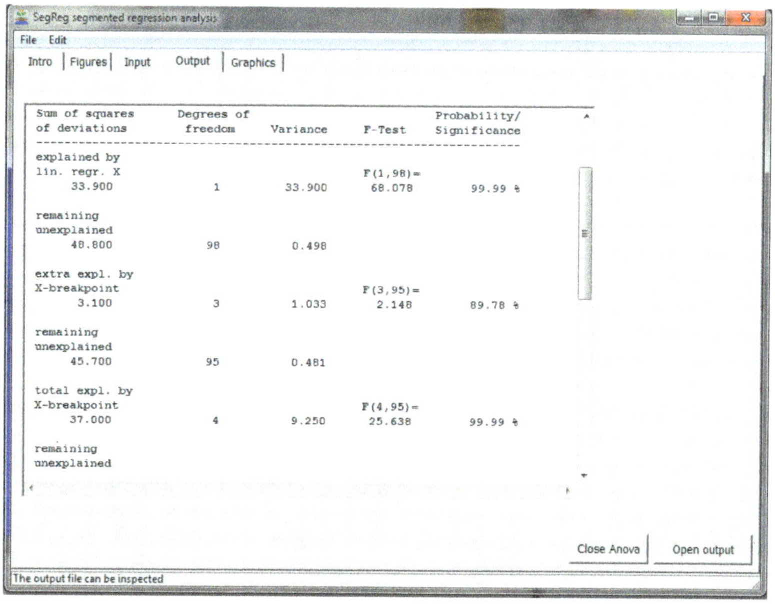

Sum of squares of deviations	Degrees of freedom	Variance	F-Test	Probability/ Significance
explained by lin. regr. X 33.900	1	33.900	F(1,98)= 68.078	99.99 %
remaining unexplained 48.800	98	0.498		
extra expl. by X-breakpoint 3.100	3	1.033	F(3,95)= 2.148	89.78 %
remaining unexplained 45.700	95	0.481		
total expl. by X-breakpoint 37.000	4	9.250	F(4,95)= 25.638	99.99 %
remaining unexplained				

Example of an Anova table: in this case the introduction of a break point is barely significant.

In addition, use is made of the correlation coefficient of all data (Ra), the coefficient of determination or coefficient of explanation, confidence intervals of the regression functions, and Anova analysis.

The coefficient of determination for all data (Cd), that is to be maximized under the conditions set by the significance tests, is found from:

$$C_d = 1 - \frac{\sum(y - Y_r)^2}{\sum(y - Y_a)^2}$$

where Yr is the expected (predicted) value of **y** according to the former regression equations and Ya is the average of all **y** values.

The Cd coefficient ranges between 0 (no explanation at all) to 1 (full explanation, perfect match). In a pure, unsegmented, linear regression, the values of Cd and Ra^2 are equal. In a segmented regression, Cd needs to be significantly larger than Ra^2 to justify the segmentation.

The optimal value of the breakpoint may be found such that the Cd coefficient is maximum.

No-effect range

Segmented regression is often used to detect over which range an explanatory variable (X) has no effect on the dependent variable (Y), while beyond the reach there is a clear response, be it positive or negative. The reach of no effect be found at the initial part of X domain or conversely at its last part. For the "no effect" analysis, application of the least squares method for the segmented regression analysis may not be the most appropriate technique because the aim is rather to find the longest stretch over which the Y-X relation can be considered to possess zero slope while beyond the reach the slope is significantly different from zero but knowledge about the best value of this slope is not material. The method to find the no-effect range is progressive partial regression over the range, extending the range with small steps until the regression coefficient gets significantly different from zero.

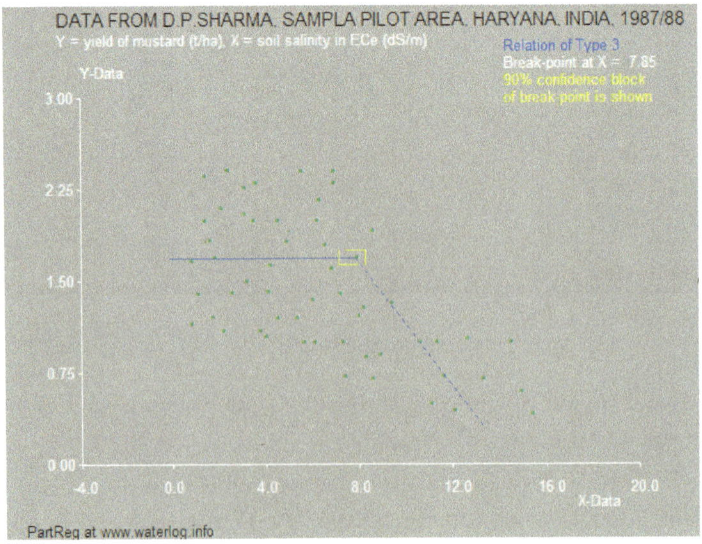

Illustration of a range from X=0 to X=7.1 over which there is no effect.

In the next figure the break point is found at X=7.9 while for the same data (see blue figure above for mustard yield), the least squares method yields a break point only at X=4.9. The latter value is lower, but the fit of the data beyond the break point is better. Hence, it will depend on the purpose of the analysis which method needs to be employed.

STEPWISE REGRESSION

In statistics, stepwise regression includes regression models in which the choice of predictive variables is carried out by an automatic procedure. Usually, this takes the form of a sequence of F-tests or t-tests, but other techniques are possible, such as adjusted R^2, Akaike information criterion, Bayesian information criterion, Mallows's C_p, PRESS, or false discovery rate.

The frequent practice of fitting the final selected model followed by reporting estimates and confidence intervals without adjusting them to take the model building process into account has led to calls to stop using stepwise model building altogether or to at least make sure model uncertainty is correctly reflected.

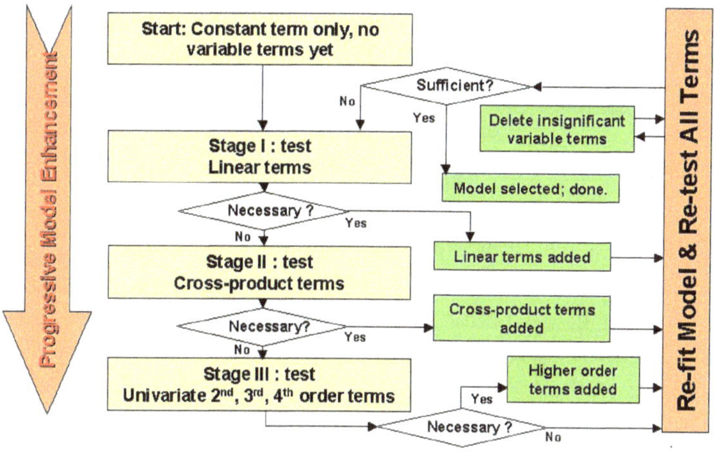

In this example from engineering, necessity and sufficiency are usually determined byF-tests. For additional consideration, when planning an experiment, computer simulation, or scientific survey to collect data for this model, one must keep in mind the number ofparameters, P, to estimate and adjust the sample size accordingly. For K variables, $P = 1_{(\text{Start})} + K_{(\text{Stage I})} + (K^2 - K)/2_{(\text{Stage II})} + 3K_{(\text{Stage III})} = 0.5K^2 + 3.5K + 1$. For $K < 17$, anefficient design of experiments exists for this type of model, a Box–Behnken design, augmented with positive and negative axial points of length $\min(2, (\text{int}(1.5 + K/4))^{1/2})$, plus point(s) at the origin. There are more efficient designs, requiring fewer runs, even for$K > 16$.

Main Approaches

The main approaches are:

- Forward selection, which involves starting with no variables in the model, testing the addition of each variable using a chosen model fit criterion, adding the variable (if any) whose inclusion gives the most statistically significant improvement of the fit, and repeating this process until none improves the model to a statistically significant extent.

- Backward elimination, which involves starting with all candidate variables, testing the deletion of each variable using a chosen model fit criterion, deleting the variable (if any) whose loss gives the most statistically insignifiant deterioration of the model fit, and repeating this process until no further variables can be deleted without a statistically significant loss of fit.
- Bidirectional elimination, a combination of the above, testing at each step for variables to be included or excluded.

Selection Criterion

A widely used algorithm was first proposed by Efroymson (1960). This is an automatic procedure for statistical *model selection* in cases where there is a large number of potential explanatory variables, and no underlying theory on which to base the model selection. The procedure is used primarily in regression analysis, though the basic approach is applicable in many forms of model selection. This is a variation on forward selection. At each stage in the process, after a new variable is added, a test is made to check if some variables can be deleted without appreciably increasing the residual sum of squares (RSS). The procedure terminates when the measure is (locally) maximized, or when the available improvement falls below some critical value.

One of the main issues with stepwise regression is that it searches a large space of possible models. Hence it is prone to overfitting the data. In other words, stepwise regression will often fit much better in sample than it does on new out-of-sample data. This problem can be mitigated if the criterion for adding (or deleting) a variable is stiff enough. The key line in the sand is at what can be thought of as the Bonferroni point: namely how significant the best spurious variable should be based on chance alone. On a t-statistic scale, this occurs at about $\sqrt{2\log p}$, where p is the number of predictors. Unfortunately, this means that many variables which actually carry signal will not be included. This fence turns out to be the right trade-off between over-fitting and missing signal. If we look at the risk of different cutoffs, then using this bound will be within a $2\log p$ factor of the best possible risk. Any other cutoff will end up having a larger such risk inflation.

Model Accuracy

A way to test for errors in models created by step-wise regression, is to not rely on the model's F-statistic, significance, or multiple R, but instead assess the model against a set of data that was not used to create the model. This is often done by building a model based on a sample of the dataset available (e.g., 70%) and use the remaining 30% dataset to assess the accuracy of the model. Accuracy is then often measured as the actual standard error (SE), MAPE, or mean error between the predicted value and the actual value in the hold-out sample. This method is particularly valuable when data are collected in different settings (e.g., different times, social vs. solitary situations) or when models are assumed to be generalizable.

Criticism

Stepwise regression procedures are used in data mining, but are controversial. Several points of criticism have been made.

- The tests themselves are biased, since they are based on the same data. Wilkinson and Dallal (1981) computed percentage points of the multiple correlation coefficient by simulation and showed that a final regression obtained by forward selection, said by the F-procedure to be significant at 0.1%, was in fact only significant at 5%.
- When estimating the degrees of freedom, the number of the candidate independent variables from the best fit selected is smaller than the total number of final model variables, causing the fit to appear better than it is when adjusting the r^2 value for the number of degrees of freedom. It is important to consider how many degrees of freedom have been used in the entire model, not just count the number of independent variables in the resulting fit.
- Models that are created may be over-simplifications of the real models of the data.

Such criticisms, based upon limitations of the relationship between a model and procedure and data set used to fit it, are usually addressed by verifying the model on an independent data set, as in the PRESS procedure.

Critics regard the procedure as a paradigmatic example of data dredging, intense computation often being an inadequate substitute for subject area expertise. Additionally, the results of stepwise regression are often used incorrectly without adjusting them for the occurrence of model selection. Especially the practice of fitting the final selected model as if no model selection had taken place and reporting of estimates and confidence intervals as if least-squares theory were valid for them, has been described as a scandal. Widespread incorrect usage and the availability of alternatives such as ensemble learning, leaving all variables in the model, or using expert judgment to identify relevant variables have led to calls to totally avoid stepwise model selection.

REFERENCES

1. Necessary Condition Analysis
2. *Armstrong, J. Scott (2012). "Illusions in Regression Analysis". International Journal of Forecasting (forthcoming). **28** (3): 689. doi:10.1016/j.ijforecast.2012.02.001.*
3. David A. Freedman, *Statistical Models: Theory and Practice*, Cambridge University Press (2005)
4. R. Dennis Cook; Sanford Weisberg Criticism and Influence Analysis in Regression,*Sociological Methodology*, Vol. 13. (1982), pp. 313–361

5. *Christopher M. Bishop (2006). Pattern Recognition and Machine Learning. Springer. p. 3.Cases [...] in which the aim is to assign each input vector to one of a finite number of discrete categories, are called classification problems. If the desired output consists of one or more continuous variables, then the task is called regression.*

6. *Waegeman, Willem; De Baets, Bernard; Boullart, Luc (2008). "ROC analysis in ordinal regression learning". Pattern Recognition Letters.* ***29****: 1–9.doi:10.1016/j. patrec.2007.07.019.*

7. A.M. Legendre. *Nouvelles méthodes pour la détermination des orbites des comètes*, Firmin Didot, Paris, 1805. "Sur la Méthode des moindres quarrés" appears as an appendix.

8. C.F. Gauss. *Theoria Motus Corporum Coelestium in Sectionibus Conicis Solem Ambientum*. (1809)

9. C.F. Gauss. *Theoria combinationis observationum erroribus minimis obnoxiae*. (1821/1823)

10. *Mogull, Robert G. (2004). Second-Semester Applied Statistics. Kendall/Hunt Publishing Company. p. 59. ISBN 0-7575-1181-3.*

11. *Galton, Francis (1989). "Kinship and Correlation (reprinted 1989)". Statistical Science. Institute of Mathematical Statistics.* ***4*** *(2): 80–86. doi:10.1214/ ss/1177012581.JSTOR 2245330.*

12. Francis Galton. "Typical laws of heredity", Nature 15 (1877), 492–495, 512–514, 532–533. *(Galton uses the term "reversion" in this paper, which discusses the size of peas.)*

13. Francis Galton. Presidential address, Section H, Anthropology. (1885) *(Galton uses the term "regression" in this paper, which discusses the height of humans.)*

14. *Yule, G. Udny (1897). "On the Theory of Correlation". Journal of the Royal Statistical Society. Blackwell Publishing.* ***60*** *(4): 812–54. doi:10.2307/2979746. JSTOR 2979746.*

15. *Pearson, Karl; Yule, G.U.; Blanchard, Norman; Lee,Alice (1903). "The Law of Ancestral Heredity". Biometrika. Biometrika Trust.* ***2*** *(2): 211–236. doi:10.1093/ biomet/2.2.211.JSTOR 2331683.*

16. *Fisher, R.A. (1922). "The goodness of fit of regression formulae, and the distribution of regression coefficients". Journal of the Royal Statistical Society. Blackwell Publishing.* ***85****(4): 597–612. doi:10.2307/2341124. JSTOR 2341124.*

17. *Ronald A. Fisher (1954). Statistical Methods for Research Workers (Twelfth ed.).Edinburgh: Oliver and Boyd. ISBN 0-05-002170-2.*

18. *Aldrich, John (2005). "Fisher and Regression". Statistical Science.* ***20*** *(4): 401–417.doi:10.1214/088342305000000331. JSTOR 20061201.*

19. Rodney Ramcharan. Regressions: Why Are Economists Obessessed with Them?March 2006. Accessed 2011-12-03.

20. N. Cressie (1996) Change of Support and the Modiable Areal Unit Problem. Geographical Systems 3:159–180.

21. *Fotheringham, A. Stewart; Brunsdon, Chris; Charlton, Martin (2002). Geographically weighted regression: the analysis of spatially varying relationships (Reprint ed.). Chichester, England: John Wiley. ISBN 978-0-471-49616-8.*

22. *Fotheringham, AS; Wong, DWS (1 January 1991). "The modifiable areal unit problem in multivariate statistical analysis". Environment and Planning A.* ***23*** *(7): 1025–1044.doi:10.1068/a231025.*

23. M. H. Kutner, C. J. Nachtsheim, and J. Neter (2004), "Applied Linear Regression Models", 4th ed., McGraw-Hill/Irwin, Boston (p. 25)

24. N. Ravishankar and D. K. Dey (2002), "A First Course in Linear Model Theory", Chapman and Hall/CRC, Boca Raton (p. 101)

25. Steel, R.G.D, and Torrie, J. H., *Principles and Procedures of Statistics with Special Reference to the Biological Sciences.*, McGraw Hill, 1960, page 288.

26. Chiang, C.L, (2003) *Statistical methods of analysis*, World Scientific. ISBN 981-238-310-7 - page 274 section 9.7.4 "interpolation vs extrapolation"

27. *Good, P. I.; Hardin, J. W. (2009). Common Errors in Statistics (And How to Avoid Them)(3rd ed.). Hoboken, New Jersey: Wiley. p. 211. ISBN 978-0-470-45798-6.*

28. *Tofallis, C. (2009). "Least Squares Percentage Regression". Journal of Modern Applied Statistical Methods. 7: 526–534. doi:10.2139/ssrn.1406472.*

29. *YangJing Long (2009). "Human age estimation by metric learning for regression problems" (PDF). Proc. International Conference on Computer Analysis of Images and Patterns: 74–82.*

CHAPTER 7

Categorical Variable

Chapter Contents

- **Introduction**
- **Examples of Categorical Variables**
- **Number of Possible Values**
- **Categorical Variables and Regression**

A categorical or discrete variable is one that has two or more categories (values). There are two types of categorical variable, nominal and ordinal. A nominal variable has no intrinsic ordering to its categories. For example, gender is a categorical variable having two categories (male and female) with no intrinsic ordering to the categories. An ordinal variable has a clear ordering. For example, temperature as a variable with three orderly categories (low, medium and high). A frequency table is a way of counting how often each category of the variable in question occurs. It may be enhanced by the addition of percentages that fall into each category.

INTRODUCTION

In statistics, a categorical variable is a variable that can take on one of a limited, and usually fixed, number of possible values, assigning each individual or other unit of observation to a particular group or nominal category on the basis of some qualitative property. In computer science and some branches of mathematics, categorical variables are referred to as enumerations or enumerated types. Commonly (though not in this article), each of the possible values of a categorical variable is referred to as a level. The probability distribution associated with a random categorical variable is called a categorical distribution. Categorical data is the statistical data type consisting of categorical variables or of data that has been converted into that form, for example as grouped data. More specifically, categorical data may derive from either or both of observations made of qualitative data, where the observations are summarized as counts or cross tabulations, or of quantitative data, where observations might be directly observed counts of events happening or might be counts of values that occur within given intervals. Often, purely categorical data are summarized in the form of a contingency table. However, particularly when considering data analysis, it is common to use the term "categorical data" to apply to data sets that, while containing some categorical variables, may also contain non-categorical variables.

A categorical variable that can take on exactly two values is termed a binary variable or dichotomous variable; an important special case is the Bernoulli variable. Categorical variables with more than two possible values are called polytomous variables; categorical variables are often assumed to be polytomous unless otherwise specified. Discretization is treating continuous data as if it were categorical. Dichotomization is treating continuous data or polytomous variables as if they were binary variables. Regression analysis often treats category membership as a quantitative dummy variable. In a simplified version, Categorical variables are data that can be seen in a graph.

EXAMPLES OF CATEGORICAL VARIABLES

Examples of values that might be represented in a categorical variable:

- The blood type of a person: A, B, AB or O.
- The state that a resident of the United States lives in.
- The political party that a voter in a European country might vote for: Christian Democrat, Social Democrat, Green Party, etc.
- The type of a rock: igneous, sedimentary or metamorphic.
- The identity of a particular word (e.g., in a language model): One of V possible choices, for a vocabulary of size V.

Notation

For ease in statistical processing, categorical variables may be assigned numeric indices, e.g. 1 through K for a K-way categorical variable (i.e. a variable that can express exactly K possible values). In general, however, the numbers are arbitrary, and have no significance beyond simply providing a convenient label for a particular value. In other words, the values in a categorical variable exist on a nominal scale: they each represent a logically separate concept, cannot necessarily be meaningfully ordered, and cannot be otherwise manipulated as numbers could be. Instead, valid operations are equivalence, set membership, and other set-related operations.

As a result, the central tendency of a set of categorical variables is given by its mode; neither the mean nor the median can be defined. As an example, given a set of people, we can consider the set of categorical variables corresponding to their last names. We can consider operations such as equivalence (whether two people have the same last name), set membership (whether a person has a name in a given list), counting (how many people have a given last name), or finding the mode (which name occurs most often). However, we cannot meaningfully compute the "sum" of Smith + Johnson, or ask whether Smith is "less than" or "greater than" Johnson. As a result, we cannot meaningfully ask what the "average name" (the mean) or the "middle-most name" (the median) is in a set of names.

Note that this ignores the concept of alphabetical order, which is a property that is not inherent in the names themselves, but in the way we construct the labels. For example, if we write the names in Cyrillic and consider the Cyrillic ordering of letters, we might get a different result of evaluating "Smith < Johnson" than if we write the names in the standard Latin alphabet; and if we write the names in Chinese characters, we cannot meaningfully evaluate "Smith < Johnson" at all, because no consistent ordering is defined for such characters. However, if we do consider the names as written, e.g., in the Latin alphabet, and define an ordering corresponding to standard alphabetical order, then we have effectively converted them into ordinal variables defined on an ordinal scale.

NUMBER OF POSSIBLE VALUES

Categorical random variables are normally described statistically by a categorical distribution, which allows an arbitrary K-way categorical variable to be expressed with separate probabilities specified for each of the K possible outcomes. Such multiple-category categorical variables are often analyzed using a multinomial distribution, which counts the frequency of each possible combination of numbers of occurrences of the various categories. Regression analysis on categorical outcomes is accomplished through multinomial logistic regression, multinomial probit or a related type of discrete choice model.

Categorical variables that have only two possible outcomes (e.g., "yes" vs. "no" or "success" vs. "failure") are known as *binary variables* (or *Bernoulli variables*). Because of their importance, these variables are often considered a separate category, with a separate distribution (the Bernoulli distribution) and separate regression models (logistic regression, probit regression, etc.). As a result, the term "categorical variable" is often reserved for cases with 3 or more outcomes, sometimes termed a *multi-way* variable in opposition to a binary variable.

It is also possible to consider categorical variables where the number of categories is not fixed in advance. As an example, for a categorical variable describing a particular word, we might not know in advance the size of the vocabulary, and we would like to allow for the possibility of encountering words that we haven't already seen. Standard statistical models, such as those involving the categorical distribution and multinomial logistic regression, assume that the number of categories is known in advance, and changing the number of categories on the fly is tricky. In such cases, more advanced techniques must be used. An example is the Dirichlet process, which falls in the realm of nonparametric statistics. In such a case, it is logically assumed that an infinite number of categories exist, but at any one time most of them (in fact, all but a finite number) have never been seen. All formulas are phrased in terms of the number of categories actually seen so far rather than the (infinite) total number of potential categories in existence, and methods are created for incremental updating of statistical distributions, including adding "new" categories.

CATEGORICAL VARIABLES AND REGRESSION

Categorical variables represent a qualitative method of scoring data (i.e. represents categories or group membership). These can be included as independent variables in a regression analysis or as dependent variables in logistic regression or probit regression, but must be converted to quantitative data in order to be able to analyze the data. One does so through the use of coding systems. Analyses are conducted such that only g -1 (g being the number of groups) are coded. This minimizes redundancy while still representing the complete data set as no additional information would be gained from coding the total g groups: for example, when coding gender (where g = 2: male and female), if we only code females everyone left over would necessarily be males. In general, the group that one does not code for is the group of least interest.

There are three main coding systems typically used in the analysis of categorical variables in regression: dummy coding, effects coding, and contrast coding. The regression equation takes the form of $Y = bX + a$, where b is the slope and gives the weight empirically assigned to an explanator, X is the explanatory variable, and a is the Y-intercept, and these values take on different meanings based on the coding system used. The choice of coding system does not affect the F or R^2 statistics. However, one chooses a coding system based on the comparison of interest since the interpretation of b values will vary.

Dummy Coding

Dummy coding is used when there is a control or comparison group in mind. One is therefore analyzing the data of one group in relation to the comparison group: a represents the mean of the control group and b is the difference between the mean of the experimental group and the mean of the control group. It is suggested that three criteria be met for specifying a suitable control group: the group should be a well-established group (e.g. should not be an "other" category), there should be a logical reason for selecting this group as a comparison (e.g. the group is anticipated to score highest on the dependent variable), and finally, the group's sample size should be substantive and not small compared to the other groups.

In dummy coding, the reference group is assigned a value of 0 for each code variable, the group of interest for comparison to the reference group is assigned a value of 1 for its specified code variable, while all other groups are assigned 0 for that particular code variable.

The b values should be interpreted such that the experimental group is being compared against the control group. Therefore, yielding a negative b value would entail the experimental group have scored less than the control group on the dependent variable. To illustrate this, suppose that we are measuring optimism among several nationalities and we have decided that French people would serve as a useful control. If we are comparing them against Italians, and we observe a negative b value, this would suggest Italians obtain lower optimism scores on average.

The following table is an example of dummy coding with *French* as the control group and C1, C2, and C3 respectively being the codes for *Italian*, *German*, and *Other* (neither French nor Italian nor German):

Nationality	C1	C2	C3
French	0	0	0
Italian	1	0	0
German	0	1	0
Other	0	0	1

Effects Coding

In the effects coding system, data are analyzed through comparing one group to all other groups. Unlike dummy coding, there is no control group. Rather, the comparison is being made at the mean of all groups combined (a is now the grand mean). Therefore, one is not looking for data in relation to another group but rather, one is seeking data in relation to the grand mean.

Effects coding can either be weighted or unweighted. Weighted effects coding is simply calculating a weighted grand mean, thus taking into account the sample size in each variable. This is most appropriate in situations where the sample is representative of the population in question. Unweighted effects coding is most appropriate in situations where differences in sample size are the result of incidental factors. The interpretation of b is different for each: in unweighted effects coding b is the difference between the mean of the experimental group and the grand mean, whereas in the weighted situation it is the mean of the experimental group minus the weighted grand mean.

In effects coding, we code the group of interest with a 1, just as we would for dummy coding. The principal difference is that we code -1 for the group we are least interested in. Since we continue to use a g - 1 coding scheme, it is in fact the -1 coded group that will not produce data, hence the fact that we are least interested in that group. A code of 0 is assigned to all other groups.

The b values should be interpreted such that the experimental group is being compared against the mean of all groups combined (or weighted grand mean in the case of weighted effects coding). Therefore, yielding a negative b value would entail the coded group as having scored less than the mean of all groups on the dependent variable. Using our previous example of optimism scores among nationalities, if the group of interest is Italians, observing a negative b value suggest they obtain a lower optimism score.

The following table is an example of effects coding with *Other* as the group of least interest.

Nationality	C1	C2	C3
French	0	0	1
Italian	1	0	0

German	0	1	0
Other	−1	−1	−1

Contrast Coding

The contrast coding system allows a researcher to directly ask specific questions. Rather than having the coding system dictate the comparison being made (i.e., against a control group as in dummy coding, or against all groups as in effects coding) one can design a unique comparison catering to one's specific research question. This tailored hypothesis is generally based on previous theory and/or research. The hypotheses proposed are generally as follows. first, there is the central hypothesis which postulates a large difference between two sets of groups; the second hypothesis suggests that within each set, the differences among the groups are small. Through its a priori focused hypotheses, contrast coding may yield an increase in power of the statistical test when compared with the less directed previous coding systems.

Certain differences emerge when we compare our a priori coefficients between ANOVA and regression. Unlike when used in ANOVA, where it is at the researcher's discretion whether they choose coefficient values that are either orthogonal or non-orthogonal, in regression, it is essential that the coefficient values assigned in contrast coding be orthogonal. Furthermore, in regression, coefficient values must be either in fractional or decimal form. They cannot take on interval values.

The construction of contrast codes is restricted by three rules:

- The sum of the contrast coefficients per each code variable must equal zero.
- The difference between the sum of the positive coefficients and the sum of the negative coefficients should equal 1.
- Coded variables should be orthogonal.

Violating rule 2 produces accurate R^2 and F values, indicating that we would reach the same conclusions about whether or not there is a significant difference; however, we can no longer interpret the b values as a mean difference.

To illustrate the construction of contrast codes consider the following table. Coefficients were chosen to illustrate our a priori hypotheses: Hypothesis 1: French and Italian persons will score higher on optimism than Germans (French = +0.33, Italian = +0.33, German = −0.66). This is illustrated through assigning the same coefficient to the French and Italian categories and a different one to the Germans. The signs assigned indicate the direction of the relationship (hence giving Germans a negative sign is indicative of their lower hypothesized optimism scores). Hypothesis 2: French and Italians are expected to differ on their optimism scores (French = +0.50, Italian = −0.50, German = 0). Here, assigning a zero value to Germans demonstrates their non-inclusion in the analysis of this hypothesis. Again, the signs assigned are indicative of the proposed relationship.

Nationality	**C1**	**C2**
French	+0.33	+0.50

Italian	+0.33	−0.50
German	−0.66	0

Nonsense Coding

Nonsense coding occurs when one uses arbitrary values in place of the designated "0"s "1"s and "-1"s seen in the previous coding systems. Although it produces correct mean values for the variables, the use of nonsense coding is not recommended as it will lead to uninterpretable statistical results.

Interactions

An interaction may arise when considering the relationship among three or more variables, and describes a situation in which the simultaneous influence of two variables on a third is not additive. Interactions may arise with categorical variables in two ways: either categorical by categorical variable interactions, or categorical by continuous variable interactions.

Categorical by Categorical Variable Interactions

This type of interaction arises when we have two categorical variables. In order to probe this type of interaction, one would code using the system that addresses the researcher's hypothesis most appropriately. The product of the codes yields the interaction. One may then calculate the *b* value and determine whether the interaction is significant.

Categorical by Continuous Variable Interactions

Simple slopes analysis is a common post hoc test used in regression which is similar to the simple effects analysis in ANOVA, used to analyze interactions. In this test, we are examining the simple slopes of one independent variable at specific values of the other independent variable. Such a test is not limited to use with continuous variables, but may also be employed when the independent variable is categorical. We cannot simply choose values to probe the interaction as we would in the continuous variable case because of the nominal nature of the data (i.e., in the continuous case, one could analyze the data at high, moderate, and low levels assigning 1 standard deviation above the mean, at the mean, and at one standard deviation below the mean respectively). In our categorical case we would use a simple regression equation for each group to investigate the simple slopes. It is common practice to standardize or center variables to make the data more interpretable in simple slopes analysis; however, categorical variables should never be standardized or centered. This test can be used with all coding systems.

REFERENCES

1. Yates, Daniel S.; Moore, David S; Starnes, Daren S. (2003). The Practice of Statistics (2nd ed.). New York: Freeman. ISBN 978-0-7167-4773-4.
2. Cohen, J.; Cohen, P.; West, S. G.; & Aiken, L. S. (2003). Applied multiple regression/correlation analysis for the behavioural sciences (3rd ed.). New York, NY: Routledge.
3. Hardy, Melissa (1993). Regression with dummy variables. Newbury Park, CA: Sage

CHAPTER 8

Survival Analysis

Chapter Contents

- **Introduction**
- **Introduction To Survival Analysis**
- **Kaplan-Meier Plot For The Aml Data**
- **Cox Proportional Hazards (Ph) Regression Analysis**
- **Tree-Structured Survival Models**
- **General Formulation**
- **Censoring**
- **Non-Parametric Estimation**

Survival analysis is generally defined as a set of methods for analyzing data where the outcome variable is the time until the occurrence of an event of interest. The event can be death, occurrence of a disease, marriage, divorce, etc. The time to event or survival time can be measured in days, weeks, years, etc. For example, if the event of interest is heart attack, then the survival time can be the time in years until a person develops a heart attack.

In survival analysis, subjects are usually followed over a specified time period and the focus is on the time at which the event of interest occurs. Why not use linear regression to model the survival time as a function of a set of predictor variables? First, survival times are typically positive numbers; ordinary linear regression may not be the best choice unless these times are first transformed in a way that removes this restriction. Second, and more importantly, ordinary linear regression cannot effectively handle the censoring of observations.

INTRODUCTION

Survival analysis is a branch of statistics for analyzing the expected duration of time until one or more events happen, such as death in biological organisms and failure in mechanical systems. This topic is called reliability theory or reliability analysis in engineering, duration analysis or duration modelling in economics, and event history analysis in sociology. Survival analysis attempts to answer questions such as: what is the proportion of a population which will survive past a certain time? Of those that survive, at what rate will they die or fail? Can multiple causes of death or failure be taken into account? How do particular circumstances or characteristics increase or decrease the probability of survival?

To answer such questions, it is necessary to define "lifetime". In the case of biological survival, death is unambiguous, but for mechanical reliability, failure may not be well-defined, for there may well be mechanical systems in which failure is partial, a matter of degree, or not otherwise localized in time. Even in biological problems, some events (for example, heart attack or other organ failure) may have the same ambiguity. The theory outlined below assumes well defined events at specific times; other cases may be better treated by models which explicitly account for ambiguous events.

More generally, survival analysis involves the modelling of time to event data; in this context, death or failure is considered an "event" in the survival analysis literature – traditionally only a single event occurs for each subject, after which the organism or mechanism is dead or broken. *Recurring event* or *repeated event* models relax that assumption. The study of recurring events is relevant in systems reliability, and in many areas of social sciences and medical research.

INTRODUCTION TO SURVIVAL ANALYSIS

Survival analysis is used in several ways:

- To describe the survival times of members of a group
    - Life tables
    - Kaplan-Meier curves
    - Survival function
    - Hazard function
- To compare the survival times of two or more groups
    - Log-rank test
- To describe the effect of categorical or quantitative variables on survival
    - Cox proportional hazards regression
    - Parametric survival models
    - Survival trees
    - Survival random forests

Definitions of Common Terms in Survival Analysis

The following terms are commonly used in survival analyses.

- Event: Death, disease occurrence, disease recurrence, recovery, or other experience of interest
- Time: The time from the beginning of an observation period (such as surgery or beginning treatment) to (i) an event, or (ii) end of the study, or (iii) loss of contact or withdrawal from the study.
- Censoring / Censored observation: If a subject does not have an event during the observation time, they are described as censored. The subject is censored in the sense that nothing is observed or known about that subject after the time of censoring. A censored subject may or may not have an event after the end of observation time.
- Survival function S(t): The probability that a subject survives longer than time t.

Example: Acute Myelogenous Leukemia Survival Data

This example uses the Acute Myelogenous Leukemia survival data set "aml" from the "survival" package in R. The data set is from Miller (1997) The question at the time was whether the standard course of chemotherapy should be extended ('maintained') for additional cycles.

- Time is indicated by the variable "time", which is the survival or censoring time
- Event (recurrence of aml cancer) is indicated by the variable "status". 0 = no event (censored), 1 = event (recurrence)
- Treatment group: the variable "x" indicates if maintenance chemotherapy was given

The last observation (11), at 161 weeks, is censored. Censoring indicates that the patient did not have an event (no recurrence of aml cancer). Another subject, observation 3, was censored at 13 weeks (indicated by status=0). This subject was only in the study for 13 weeks, and the aml cancer did not recur during those 13 weeks. It is possible that this patient was enrolled near the end of the study, so that they could only be observed for 13 weeks. It is also possible that the patient was enrolled early in the study, but was lost to follow up or withdrew from the study. The table shows that other subjects were censored at 16, 28, and 45 weeks (observations 17, 6, and 9 with status=0). The remaining subjects all experienced events (recurrence of aml cancer) while in the study. The question of interest is whether recurrence occurs later in maintained patients than in non-maintained patients.

KAPLAN-MEIER PLOT FOR THE AML DATA

The Survival function S(t), is the probability that a subject survives longer than time t. S(t) is theoretically a smooth curve, but it is usually estimated using the Kaplan-Meier(KM) curve. The graph shows the KM plot for the aml data.

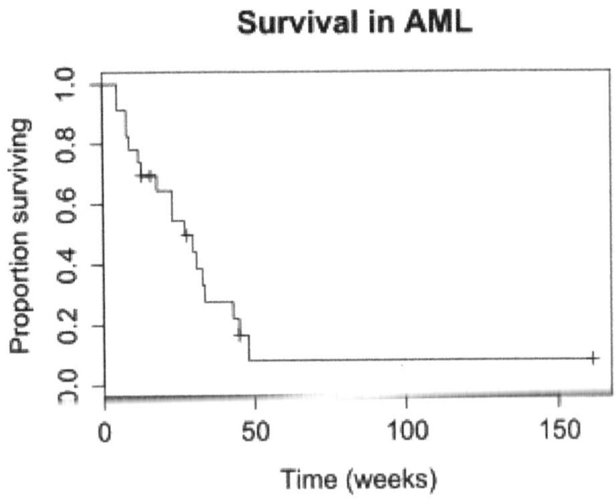

Kaplan-Meier plot of AML survival data set

The KM plot is interpreted as follows.

- The x-axis is time, from zero (when observation began) to the last observed time point.
- The y axis is the proportion of subjects surviving. At time zero, 100% of the subjects are alive without an event.
- The solid line (similar to a staircase) shows the events.
- A vertical drop indicates an event. In the aml table shown above, two subjects had events at 5 weeks, two had events at 8 weeks, one had an event at 9 weeks, and so on. These events at 5 weeks, 8 weeks and so on are indicated by the vertical drops in the KM plot at those time points.
- At the far right end of the KM plot there is a tick mark at 161 weeks. The vertical tick mark indicates that a patient was censored at this time. In the aml data table five subjects were censored, at 13, 16, 28, 45 and 161 weeks. There are five tick marks in the KM plot, corresponding to these censored observations.

Life Table for the aml Data

A Life tables summarizes survival data in terms of the number of events and the proportion surviving at each event time point. The life table for the aml data, created using the R software, is shown.

time	n.risk	n.event	survival	std.err	lower 95% CI	upper 95% CI
5	23	2	0.913	0.0588	0.8049	1
8	21	2	0.8261	0.079	0.6848	0.996
9	19	1	0.7826	0.086	0.631	0.971
12	18	1	0.7391	0.0916	0.5798	0.942
13	17	1	0.6957	0.0959	0.5309	0.912
18	14	1	0.646	0.1011	0.4753	0.878
23	13	2	0.5466	0.1073	0.3721	0.803
27	11	1	0.4969	0.1084	0.324	0.762
30	9	1	0.4417	0.1095	0.2717	0.718
31	8	1	0.3865	0.1089	0.2225	0.671
33	7	1	0.3313	0.1064	0.1765	0.622
34	6	1	0.2761	0.102	0.1338	0.569
43	5	1	0.2208	0.0954	0.0947	0.515
45	4	1	0.1656	0.086	0.0598	0.458
48	2	1	0.0828	0.0727	0.0148	0.462

Life table for the aml data

The life table summarizes the events and the proportion surviving at each event time point. The columns in the life table have the following interpretation.

- time gives the time points at which events occur.
- n.risk is the number of subjects at risk immediately before the time point, t. Being "at risk" means that the subject has not had an event before time t, and is not censored before or at time t.
- n.event is the number of subjects who have events at time t.
- survival is the proportion surviving, as determined using the Kaplan-Meier product-limit estimate.
- std.err is the standard error of the estimated survival. The standard error of the Kaplan-Meier product-limit estimate at time it is calculated using Greenwood's formula, and depends on the number at risk (n.risk in the table), the number of deaths (n.event in the table) and the proportion surviving (survival in the table).
- lower 95% CI and upper 95% CI are the lower and upper 95% confidence bounds for the proportion surviving.

Log-rank test: Testing for Differences in Survival in the aml data

The logrank test compares the survival times of two or more groups. This example uses a logrank test for a difference in survival in the maintained versus non-maintained treatment groups in the aml data. The graph shows KM plots for the aml data broken out by treatment group, which is indicated by the variable "x" in the data.

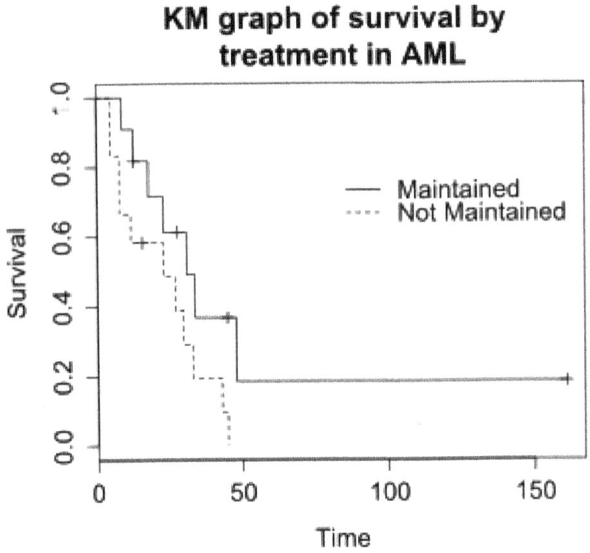

Kaplan-Meier graph by treatment group in aml

The null hypothesis for a logrank test is that the groups have the same survival. The expected number of subjects surviving at each time point in each is adjusted for the number of subjects at risk in the groups at each event time. The logrank test determines if the observed number of events in each group is significantly different from the expected number. The formal test is based on a chi-squared statistic. When the log rank statistic is large, it is evidence for a different in the survival times between the groups. The log rank statistic has a chi-squared distribution with one degree of freedom, and the p-value is calculated using the chi-squared distribution.

The log rank test for difference in survival gives a p-value of p=0.0653, indicating that the treatment groups do not differ significantly in survival, although they approach significance. The sample size of 23 subjects is modest, so there is little power to detect differences between the treatment groups. The chi-squared test is based on asymptotic approximation, so the p-value should be regarded with caution for small sample sizes.

COX PROPORTIONAL HAZARDS (PH) REGRESSION ANALYSIS

Kaplan-Meier curves and logrank tests are most useful when the predictor variable is categorical (e.g., drug vs. placebo), or takes a small number of values (e.g., drug doses 0, 20, 50, and 100 mg/day) that can be treated as categorical. The logrank test and K-M curves don't work easily with quantitative predictors such as gene expression, white blood count, or age. For quantitative predictor variables, an alternative method is Cox proportional hazards regression analysis. Cox PH models work also with categorical predictor variables, which are encoded as {0,1} indicator or dummy variables. The logrank test is a special case of a Cox PH analysis, and can be performed using Cox PH software.

Example: Cox Proportional Hazards Regression Analysis for Melanoma

This example uses the melanoma data set from Dalgaard Chapter 12.

Data are in the R package ISwR. The Cox proportional hazards regression using R gives the results shown in the box.

The Cox regression results are interpreted as follows.

- Sex is encoded as a numeric vector. 1: female, 2: male. The R summary for the cox model gives the hazard ratio (HR) for the second group relative to the first group, that is, male versus female.
- coef= 0.662 is the estimated logarithm of the hazard ratio for males versus females.
- exp(coef) = 1.94 = exp(0.662) The log of the hazard ratio (coef= 0.662) is transformed to the hazard ratio using exp(coef). The summary for the Cox model gives the hazard ratio for the second group relative to the first group, that is, male versus female. The estimated hazard ratio = 1.94 indicates that males have higher risk of death (lower survival rates) than females, in these data.
- se(coef) = 0.265 is the standard error of the log hazard ratio.
- z = 2.5 = coef/se(coef) = 0.662/0.265. Dividing the coef by its standard error gives the z score:
- p=0.013. The p-value corresponding to z=2.5 for sex is p=0.013, indicating that there is a significant difference in survival as a function of sex.

The summary output also gives upper and lower 95% confidence intervals for the hazard ratio, lower 95% bound = 1.15, upper 95% bound = 3.26.

Finally, the output gives p-values for three alternative tests for overall significance of the model:

- Likelihood ratio test= 6.15 on 1 df, p=0.0131
- Wald test = 6.24 on 1 df, p=0.0125
- Score (logrank) test = 6.47 on 1 df, p=0.0110

These three methods are asymptotically equivalent. For large enough N, they will give similar results. For small N, they may differ somewhat. The last row, "Score (logrank) test" is the result for the logrank test, with p=0.011, the same result as the logrank test, because the logrank test is a special case of a Cox PH regression. The Likelihood ratio test has better behavior for small sample sizes, so it is generally preferred.

Cox Model Using a Covariate In The Melanoma Data

The Cox model extends the logrank test by allowing the inclusion of additional covariates. This example use the melanom data set where the predictor variables include a continuous covariate, the thickness of the tumor (variable name = "thick")

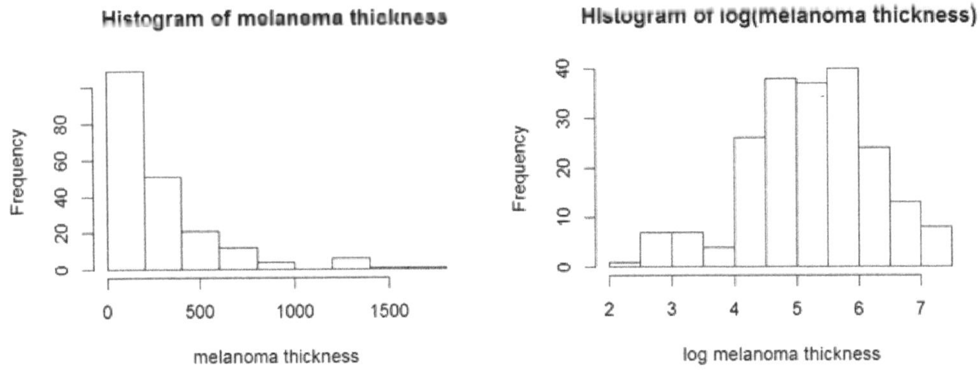

Histograms of melanoma tumor thickness

In the histograms, the thickness values don't look normally distributed. Regression models, including the Cox model, generally give more reliable results with normally-distributed variables. For this example use a log transform. The log of the thickness of the tumor looks to be more normally distributed, so the Cox models will use log thickness. The Cox PH analysis gives the results in the box.

The p-value for all three overall tests (likelihood, Wald, and score) are significant, indicating that the model is significant. The p-value for log(thick) is 6.9e-07, with a hazard ratio HR = exp(coef) = 2.18, indicating a strong relationship between the thickness of the tumor and increased risk of death.

By contrast, the p-value for sex is now p=0.088. The hazard ratio HR = exp(coef) = 1.58, with a 95% confidence interval of 0.934 to 2.68. Because the confidence interval for HR includes 1, these results indicate that sex makes a smaller contribution to the difference in the HR after controlling for the thickness of the tumor, and only trend toward significance. Examination of graphs of log(thickness) by sex and a t-test of log(thickness) by sex both indicate that there is a significant difference between men and women in the thickness of the tumor when they first see the clinician.

The Cox model assumes that the hazards are proportional. The proportional hazard assumption may be tested using the R function cox.zph(). A p-value is less than 0.05 indicates that the hazards are not proportional. For the melanoma data, p=0.222, indicating that the hazards are, at least approximately, proportional. Additional tests and graphs for examining a Cox model are described in the textbooks cited.

Extensions to Cox models

Cox models can be extended to deal with variations on the simple analysis.

- Stratification. The subjects can be divided into strata, where subjects within a stratum are expected to be relatively more similar to each other than to randomly chosen subjects from other strata. The regression parameters are assumed to be the same across the strata, but a different baseline hazard may exist for each stratum. Stratification is useful for analyses using matched subjects, for dealing with patient subsets, such as different clinics, and for dealing with violations of the proportional hazard assumption.
- Time-varying covariates. Some variables, such as gender and treatment group, generally stay the same in a clinical trial. Other clinical variables, such as serum protein levels or dose of concomitant medications may change over the course of a study. Cox models may be extended for such time-varying covariates.

TREE-STRUCTURED SURVIVAL MODELS

The Cox PH regression model is a linear model. It is similar to linear regression and logistic regression. Specifically, these methods assume that a single line, curve, plane, or surface is sufficient to separate groups (alive, dead) or to estimate a quantitative response (survival time).

In some cases alternative partitions give more accurate classification or quantitative estimates. One set of alternative methods are tree-structured survival models, including survival random forests. Tree-structured survival models may give more accurate predictions than Cox models. Examining both types of models for a given data set is a reasonable strategy.

Example Survival Tree Analysis

This example of a survival tree analysis uses the R package "rpart". The example is based on 146 stage C prostate cancer patients in the data set stagec in rpart. Rpart and the stagec example are described in the PDF document "An Introduction to Recursive Partitioning Using the RPART Routines". Terry M. Therneau, Elizabeth J. Atkinson, Mayo Foundation. September 3, 1997.

The variables in stagec are:

- pgtime time to progression, or last follow-up free of progression
- pgstat status at last follow-up (1=progressed, 0=censored)
- age age at diagnosis
- eet early endocrine therapy (1=no, 0=yes)
- ploidy diploid/tetraploid/aneuploid DNA pattern
- g2 % of cells in G2 phase

- grade tumor grade (1-4)
- gleason Gleason grade (3-10)

Each branch in the tree indicates a split on the value of a variable. For example, the root of the tree splits subjects with grade < 2.5 versus subjects with grade 2.5 or greater. The terminal nodes indicate the number of subjects in the node, the number of subjects who have events, and the relative event rate compared to the root. In the node on the far left, the values 1/33 indicate that 1 of the 33 subjects in the node had an event, and that the relative event rate is 0.122. In the node on the far right bottom, the values 11/15 indicate that 11 of 15 subjects in the node had an event, and the relative event rate is 2.7.

Survival Random Forests

An alternative to building a single survival tree is to build many survival trees, where each tree is constructed using a sample of the data, and average the trees to predict survival. This is the method underlying the survival random forest models. Survival random forest analysis is available in the R package "randomForestSRC". The randomForestSRC package includes an example survival random forest analysis using the data set pbc. This data is from the Mayo Clinic Primary Biliary Cirrhosis (PBC) trial of the liver conducted between 1974 and 1984. In the example, the random forest survival model gives more accurate predictions of survival than the Cox PH model. The prediction errors are estimated by bootstrap re-sampling.

GENERAL FORMULATION

Survival Function

The object of primary interest is the survival function, conventionally denoted S, which is defined as

$$S(t) = \Pr(T > t)$$

where t is some time, T is a random variable denoting the time of death, and "Pr" stands for probability. That is, the survival function is the probability that the time of death is later than some specified time t. The survival function is also called the *survivor function* or *survivorship function* in problems of biological survival, and the *reliability function* in mechanical survival problems. In the latter case, the reliability function is denoted $R(t)$.

Usually one assumes $S(0) = 1$, although it could be less than 1 if there is the possibility of immediate death or failure.

The survival function must be non-increasing: $S(u) \leq S(t)$ if $u \geq t$. This property follows directly because $T>u$ implies $T>t$. This reflects the notion that survival to a later age is only possible if all younger ages are attained. Given this property, the lifetime distribution function and event density (F and f below) are well-defined.

The survival function is usually assumed to approach zero as age increases without bound, i.e., $S(t) \to 0$ as $t \to \infty$, although the limit could be greater than zero if eternal life is possible. For instance, we could apply survival analysis to a mixture of stable and unstable carbon isotopes; unstable isotopes would decay sooner or later, but the stable isotopes would last indefinitely.

Lifetime Distribution Function and Event Density

Related quantities are defined in terms of the survival function.

The lifetime distribution function, conventionally denoted F, is defined as the complement of the survival function,

$$F(t) = \Pr(T \leq t) = 1 - S(t).$$

If F is differentiable then the derivative, which is the density function of the lifetime distribution, is conventionally denoted f,

$$f(t) = F'(t) = \frac{d}{dt}F(t).$$

The function f is sometimes called the **event density**; it is the rate of death or failure events per unit time.

The survival function can be expressed in terms of probability distribution and probability density functions

$$S(t) = \Pr(T > t) = \int_t^\infty f(u)\,du = 1 - F(t).$$

Similarly, a survival event density function can be defined as

$$s(t) = S'(t) = \frac{d}{dt}S(t) = \frac{d}{dt}\int_t^\infty f(u)\,du = \frac{d}{dt}[1 - F(t)] = -f(t).$$

The survival event density function in other fields, such as Statistical Physics is known as the first passage time density.

Hazard Function and Cumulative Hazard Function

The hazard function, conventionally denoted λ, is defined as the event rate at time t conditional on survival until time t or later (that is, $T \geq t$). Suppose that an item has survived for a time t and we desire the probability that it will not survive for an additional time dt. That is, consider $P\{X \in (t, t + dt) | X > t\}$

$$\lambda(t) = \lim_{dt \to 0} \frac{\Pr(t \leq T < t + dt)}{dt \cdot S(t)} = \frac{f(t)}{S(t)} = -\frac{S'(t)}{S(t)}.$$

Force of mortality is a synonym of *hazard function* which is used particularly in demography and actuarial science, where it is denoted by μ. The term *hazard rate* is another synonym.

The force of mortality of the survival function is defined as

$$\mu(x) = -\frac{d}{dx}\ln(S(x)) = \frac{f(x)}{S(x)}$$

The force of mortality is also called the force of failure. is the probability density function of the distribution.

In actuarial science, the hazard rate is the rate of death for lives aged x. For a life aged x, the force of mortality t years later is the force of mortality for a (x + t)–year old. The hazard rate is also called the failure rate. Hazard rate and failure rate are names used in reliability theory.

Any function is a hazard function if and only if it satisfies the following properties:

1. $h(x) \geq 0 \forall (x \geq 0)$,
2. $\int_0^{\infty} h(x)dx = \infty$.

In fact, the hazard rate is usually more informative about the underlying mechanism of failure than the other representatives of a lifetime distribution.

The hazard function must be non-negative, $\lambda(t) \geq 0$, and its integral over $[0, \infty]$ must be infinite, but is not otherwise constrained; it may be increasing or decreasing, non-monotonic, or discontinuous. An example is the bathtub curve hazard function, which is large for small values of t, decreasing to some minimum, and thereafter increasing again; this can model the property of some mechanical systems to either fail soon after operation, or much later, as the system ages.

The hazard function can alternatively be represented in terms of the cumulative hazard function, conventionally

$$\Lambda(t) = -\log S(t)$$

so transposing signs and exponentiating

$$S(t) = \exp(-\Lambda(t))$$

or differentiating (with the chain rule)

$$\frac{d}{dt}\Lambda(t) = -\frac{S'(t)}{S(t)} = \lambda(t).$$

The name "cumulative hazard function" is derived from the fact that

$$\Lambda(t) = \int_0^t \lambda(u)\,du$$

which is the "accumulation" of the hazard over time.

From the definition of $\Lambda(t)$, we see that it increases without bound as t tends to infinity (assuming that $S(t)$ tends to zero). This implies that $\lambda(t)$ must not decrease too quickly, since, by definition, the cumulative hazard has to diverge. For example, $\exp(-t)$ is not the hazard function of any survival distribution, because its integral converges to 1.

Quantities Derived From The Survival Distribution

Future lifetime at a given time t_0 is the time remaining until death, given survival to age t_0. Thus, it is $T - t_0$ in the present notation. The expected future lifetime is the expected value of future lifetime. The probability of death at or before age $t_0 + t$, given survival until age t_0, is just

$$P(T \leq t_0 + t \mid T > t_0) = \frac{P(t_0 < T \leq t_0 + t)}{P(T > t_0)} = \frac{F(t_0 + t) - F(t_0)}{S(t_0)}.$$

Therefore, the probability density of future lifetime is

$$\frac{d}{dt}\frac{F(t_0 + t) - F(t_0)}{S(t_0)} = \frac{f(t_0 + t)}{S(t_0)}$$

and the expected future lifetime is

$$\frac{1}{S(t_0)} \int_0^\infty t\, f(t_0 + t)\, dt = \frac{1}{S(t_0)} \int_{t_0}^\infty S(t)\, dt,$$

where the second expression is obtained using integration by parts.

For $t_0 = 0$, that is, at birth, this reduces to the expected lifetime.

In reliability problems, the expected lifetime is called the *mean time to failure*, and the expected future lifetime is called the *mean residual lifetime*.

As the probability of an individual surviving until age t or later is $S(t)$, by definition, the expected number of survivors at age t out of an initial population of n newborns is $n \times S(t)$, assuming the same survival function for all individuals. Thus the expected proportion of survivors is $S(t)$. If the survival of different individuals is independent, the number of survivors at age t has a binomial distribution with parameters n and $S(t)$, and the variance of the proportion of survivors is $S(t) \times (1-S(t))/n$.

The age at which a specified proportion of survivors remain can be found by solving the equation $S(t) = q$ for t, where q is the quantile in question. Typically one is interested in the **median lifetime**, for which $q = 1/2$, or other quantiles such as $q = 0.90$ or $q = 0.99$.

One can also make more complex inferences from the survival distribution. In mechanical reliability problems, one can bring cost (or, more generally, utility) into consideration, and thus solve problems concerning repair or replacement. This leads to the study of renewal theory and reliability theory of ageing and longevity.

CENSORING

Censoring is a form of missing data problem which is common in survival analysis. Ideally, both the birth and death dates of a subject are known, in which case the lifetime is known. If it is known only that the date of death is after some date, this is called *right censoring*. Right censoring will occur for those subjects whose birth date is known but who are still alive when they are lost to follow-up or when the study ends.

If a subject's lifetime is known to be less than a certain duration, the lifetime is said to be *left-censored*. Left censoring is usually applied when subjects in a study already have exhibited the event in question at the start of the study but information about when they first reached the event is unclear.

It may also happen that subjects with a lifetime less than some threshold may not be observed at all: this is called *truncation*. Note that truncation is different from left censoring, since for a left censored datum, we know the subject exists, but for a truncated datum, we may be completely unaware of the subject. Truncation is also common. In a so-called *delayed entry* study, subjects are not observed at all until they have reached a certain age. For example, people may not be observed until they have reached the age to enter school. Any deceased subjects in the pre-school age group would be unknown. Left truncated data are common in actuarial work for life insurance and pensions.

We generally encounter right-censored data. Left-censored data can occur when a person's survival time becomes incomplete on the left side of the follow-up period for the person. For example, in a epidemiological example, we may monitor a patient for a infectious disorder starting from the time when he or she is tested positive for the infection. Although we may know the right-hand side of the duration of interest, we may never know the exact time of exposure to the infectious agent.

Fitting Parameters to Data

Survival models can be usefully viewed as ordinary regression models in which the response variable is time. However, computing the likelihood function (needed for fitting parameters or making other kinds of inferences) is complicated by the censoring. The likelihood function for a survival model, in the presence of censored data, is formulated as follows. By definition the likelihood function is the conditional probability of the data given the parameters of the model. It is customary to assume that the data are independent given the parameters. Then the likelihood function is the product of the likelihood of each datum. It is convenient to partition the data into four categories: uncensored, left censored, right censored, and interval censored. These are denoted "unc.", "l.c.", "r.c.", and "i.c." in the equation below.

$$L(\theta) = \prod_{T_i \in unc.} \Pr(T = T_i \mid \theta) \prod_{i \in l.c.} \Pr(T < T_i \mid \theta) \prod_{i \in r.c.} \Pr(T > T_i \mid \theta) \prod_{i \in i.c.} \Pr(T_{i,l} < T < T_{i,r} \mid \theta).$$

For uncensored data, with T_i equal to the age at death, we have

$$\Pr(T = T_i \mid \theta) = f(T_i \mid \theta).$$

For left-censored data, such that the age at death is known to be less than T_i, we have

$$\Pr(T < T_i \mid \theta) = F(T_i \mid \theta) = 1 - S(T_i \mid \theta).$$

For right-censored data, such that the age at death is known to be greater than T_i, we have

$$\Pr(T > T_i \mid \theta) = 1 - F(T_i \mid \theta) = S(T_i \mid \theta).$$

For an interval censored datum, such that the age at death is known to be less than $T_{i,r}$ and greater than $T_{i,l}$, we have

$$\Pr(T_{i,l} < T < T_{i,r} \mid \theta) = S(T_{i,l} \mid \theta) - S(T_{i,r} \mid \theta).$$

An important application where interval-censored data arises is current status data, where an event Ti is known not to have occurred before an observation time and to have occurred before the next observation time.

NON-PARAMETRIC ESTIMATION

The Kaplan-Meier estimator can be used to estimate the survival function. The Nelson–Aalen estimator can be used to provide a non-parametric estimate of the cumulative hazard rate function.

Computer Software for Survival Analysis

The UCLA website http://www.ats.ucla.edu/stat/ has numerous examples of statistical analyses using SAS, R, SPSS and STATA, including survival analyses. The textbook by Kleinbaum has examples of survival analyses using SAS, R, and other packages. The textbooks by Brostrom by Dalgaard and by Tableman and Kim give examples of survival analyses using R (or using S, and which run in R).

Survival Analysis in R

The code below performs the analyses on this Wikipedia page.

Analyses using the R Package "Survival"

The examples above use the R package "survival", except for the tree analyses described below.

```
# Install and load the survival package

install.packages("survival")
library(survival)

# sort the aml data by time
aml=aml[order(aml$time),]

aml
# Create graph of length of time that each subject was in the study
with(aml, plot(time, type="h"))

# Create the life table survival object for aml
aml.survfit = survfit(Surv(time, status == 1) ~ 1, data=aml)

# Plot the Kaplan-Meier curve for aml. Don't print the confidence interval.
plot(aml.survfit, xlab = "Time (weeks)", ylab="Proportion surviving", conf.int=FALSE, main="Survival in AML")
```

```
# Create the life table for the aml data
# The functions survfit() and Surv() create a life table survival object.
# The summary() function displays the life table
# The life table object is passed to the plot() function to create the KM plot.
aml.survfit = survfit(Surv(time, status == 1) ~ 1, data=aml)

summary(aml.survfit)

# Kaplan-Meier curve for aml with the confidence bounds.
# By default, R includes the confidence interval.
plot(aml.survfit, xlab = "Time", ylab="Proportion surviving")

# Create aml life tables and KM plots broken out by treatment (x,  "Maintained" vs. "Not maintained")
surv.by.aml.rx = survfit(Surv(time, status == 1) ~ x, data = aml)

summary(surv.by.aml.rx)


# Plot KM
plot(surv.by.aml.rx, xlab = "Time", ylab="Survival",col=c("black", "red"), lty = 1:2, main="Kaplan-Meier Survival
vs. Maintenance in AML")

# Add legend
legend(100, .6, c("Maintained", "Not maintained"), lty = 1:2, col=c("black", "red"))

# Perform the log rank test using the R function survdiff().

surv.diff.aml= survdiff(Surv(time, status == 1) ~ x, data=aml)

surv.diff.aml

# Cox Proportional Hazards regression
# melanoma data set from ISwR package, described in Dalgaard Chapter 12.
# install the ISwR package and load the library into R.
# The ISwR package currently only appears to be available for older versions of R


install.packages("ISwR")

library(ISwR)

help(melanom) # description of the melanoma data

# The log rank test is a special case of the cox proportional hazard regression analysis.
# The same analysis can be performed using the R function coxph().
# melanoma example using a log-rank test.
surv.diff.sex = survdiff(Surv(days, status == 1) ~ sex, data = melanom)

surv.diff.sex

# melanoma analysis using Cox proportional hazards regression
coxph.sex = coxph(Surv(days, status == 1) ~ sex, data = melanom)

summary(coxph.sex)

# melanoma Cox analysis including covariate ulcer thickness
```

```
# Plot the thickness values and log(thickness)
hist(melanom$thick)

hist(log(melanom$thick))

# The Cox PH analysis of melanoma data including covariate log(thick)

coxph.sex.thick = coxph(Surv(days, status == 1) ~ sex + log(thick), data = melanom)

summary(coxph.sex.thick)

# Examine thickness by sex
boxplot(log(melanom$thick) ~ melanom$sex)

t.test(log(melanom$thick) ~ melanom$sex)

# Test of proportional hazards assumption
coxph.sex = coxph(Surv(days, status == 1) ~ sex, data = melanom)

cox.zph(coxph.sex)
```

Survival Tree Analysis using the rpart Package

Rpart and the example are described in the PDF document "An Introduction to Recursive Partitioning Using the RPART Routines". Terry M. Therneau, Elizabeth J. Atkinson, Mayo Foundation. September 3, 1997.

```
install.packages("rpart")
library(rpart)

head(stagec)

# Pass a survival object from Surv() to the function rpart() to perform the analysis.
fit <- rpart(Surv(pgtime, pgstat) ~ age + eet + g2 + grade + gleason + ploidy, data=stagec)

# plot the resulting tree

plot(fit, uniform=T, branch=.4, compress=T)
text(fit, use.n=T)
# The print() function provides details of the tree not shown above
print(fit)
```

Survival Random Forest Models Using the RandomForestSRC package

Note that the R package randomSurvivalForest has been replaced by the package randomForestSRC, "Random Forests for Survival, Regression and Classification". See the randomForestSRC package for documentation on running the example.

REFERENCES

1. Miller, Rupert G. (1997), Survival analysis, John Wiley & Sons, ISBN 0-471-25218-2
2. Renganathan, Vinaitheerthan (2016-03-31). "Overview of Frequentist and Bayesian approach to Survival Analysis". Applied Medical Informatics. 38 (1): 25–38. ISSN 2067-7855.
3. Dalgaard, Peter (2008), Introductory Statistics with R (Second ed.), Springer, ISBN 978-0387790534
4. "Censoring, Left and Right." International Encyclopedia of the Social Sciences, edited by William A. Darity, Jr., 2nd ed., vol. 1, Macmillan Reference USA, 2008, pp. 473-474. libraries.state.ma.us/login?gwurl=http://ic.galegroup.com/ic/uhic/ReferenceDetailsPage/ReferenceDetailsWindow?disableHighlighting=false&displayGroupName=Reference&currPage=&scanId=&query=&prodId=UHIC&search_within_results=&p=UHIC%3AWHIC&mode=view&catId=&limiter=&display-query=&displayGroups=&contentModules=&action=e&sortBy=&documentId=GALE%7CCX3045300295&windowstate=normal&activityType=&failOverType=&commentary=&source=Bookmark&u=mlin_w_amhercol&jsid=0938fef854cc86b83b5fe8a2c4bcb54b. Accessed 6 Nov. 2016.
5. Richards, S. J. (2012). "A handbook of parametric survival models for actuarial use". Scandinavian Actuarial Journal. 2012 (4): 233–257. doi:10.1080/03461238.2010.506688.
6. Singh, R.; Mukhopadhyay, K. (2011). "Survival analysis in clinical trials: Basics and must know areas". Perspect Clin Res. 2 (4): 145–148. doi:10.4103/2229-3485.86872.
7. Kleinbaum, David G.; Klein, Mitchel (2012), Survival analysis: A Self-learning text (Third ed.), Springer, ISBN 978-1441966452
8. Brostrom, Göran (2012), Event History Analysis with R (First ed.), Chapman & Hall/CRC, ISBN 978-1439831649
9. Tableman, Mara; Kim, Jong Sung (2003), Survival Analysis Using S (First ed.), Chapman and Hall/CRC, ISBN 978-1584884088

Index

A

Analysis of variance (ANOVA) 20
arithmetic mean 33, 34, 35, 36, 37, 39, 40, 41, 43, 45, 48

C

categorical variable 241, 242, 243, 247
Censoring 251, 252, 261, 266
Central Tendency 33, 64
Coefficient of Determination 76
COEFFICIENT OF DETERMINATION 198
coefficient of dispersion (CD) 51
confounding 191, 192, 193, 194, 195, 196, 197, 198
Correlation 167, 170, 173, 175, 181, 185, 187, 188, 189, 190, 191, 201, 207, 208, 209
cumulative hazard function 260

D

Data collection 69
dependence 167, 168, 169, 171, 172, 175, 187
Descriptive statistics 31, 32, 64

E

Estimation theory 137

F

false discovery rate (FDR) 142
family wise error rate (FWER) 142
fraction of unexplained variance (FUV) 201

I

independent identically distributed (IID) 16
Interquartile range (IQR) 61

L

LINEAR REGRESSION 216
LOCAL REGRESSION 221

M

Mathematical statistics 4, 5, 8, 24
Mean 35, 36, 38, 39, 40, 41, 56, 58, 61
mean-preserving spread (MPS) 62
Means 81, 85, 94
Mean square weighted deviation (MSWD) 20
Median 34, 41, 47, 48, 52, 53, 54, 56, 58, 61, 64, 66, 67
Median absolute deviation (MAD) 61
minimum description length (MDL) 134
missing at random (MAR) 98
missing completely at random (MCAR) 99
Missing Data 98, 99
Missing not at random (MNAR) 99

N

necessary condition analysis (NCA) 212
NONLINEAR REGRESSION 219
null hypothesis significance testing (NHST) 156

O

odds ratio (OR) 82

P

Pearson correlation coefficient 168, 172, 173, 174, 177, 178, 179, 180, 182, 186, 187, 190, 201
population covariance 186
population mean 72, 77, 85, 94, 116, 120, 122
Poststratification 109
probabilistic context-free grammars (PCFG) 15
Probability Distributions 5, 44

R

Regression 5, 7, 13, 20
Regression analysis 211, 212, 224, 242, 243
regression function 7
regressors 198, 203
relative risk (RR) 83
residual sum of squares (RSS) 236
risk difference (RD) 83
ROBUST REGRESSION 224

S

Sampling 102, 103, 104, 105, 106, 108, 110, 111, 112, 113, 114, 115, 118
Segmented regression 229, 234
simple random sample (SRS) 105
standard error of the mean (SEM) 116, 120
Statistic 72, 73
statistical hypothesis 138, 139, 144, 148, 155, 156
Statistical inference 6, 9, 129, 132
statistical model 129, 130, 132, 136
Statistical Population 71
Statistics 1, 2, 4, 7, 8, 9, 16, 19, 21, 24, 25, 26, 27, 28, 29
STEPWISE REGRESSION 235
subpopulation 71
Survival analysis 249, 266, ,

U

Univariate Analysis 32